The Origin Map

The Origin Map

Discovery of a Prehistoric, Megalithic, Astrophysical Map
and Sculpture of the Universe

Thomas G. Brophy, Ph.D.
with Foreword by
Robert M. Schoch, Ph.D.
and Afterword by
John Anthony West

Writers Club Press
New York   Lincoln   Shanghai

The Origin Map
Discovery of a Prehistoric, Megalithic,
Astrophysical Map and Sculpture of the Universe

Writers Club Press
an imprint of iUniverse, Inc.

For information address:
iUniverse
2021 Pine Lake Road, Suite 100
Lincoln, NE 68512
www.iuniverse.com

ISBN: 0-595-24122-0

Printed in the United States of America

ABOUT THE COVER

The background in the cover image is the spiral galaxy NGC 4414. It is a photomontage taken by the Hubble Space Telescope in 1995, by Wendy Freedman and a team of astronomers. The galaxy is about 60 million light-years away from us. It is similar shape and size to our own Milky Way Galaxy. In our galaxy, our Sun, the star that we orbit, is located near the outer part of one of the Milky Way spiral arms. The point of convergence of the two lines in the drawing represents the point where our sun would be in the Milky Way. Montaged with the galaxy photo are two images of the matching of the constellation Orion to the stones in the 7,000 years old calendar circle diagram discovered at Nabta Playa in the Sahara desert. The dates of the star to stone matching images are about 5,000 BC and about 16,000 BC.

CONTENTS

ACKNOWLEDGEMENTS

I thank my parents, Jim and June Brophy and my sisters Jayne and Terre, for their supportive curiosity. I thank Yasuhiko Kimura and Laara Lindo, leaders of the Twilight Club and the University of Science and Philosophy, and geologist Daniel Krummenacher, for their general comments and encouragement. I thank Marjorie Fox for a cover design idea. I thank Fred Wendorf for his leadership of the carefully documented excavations at Nabta Playa. I thank attorney Valerie Ann Nemeth for intellectual property consultations, and attorney Richard Warburg for patent advice. I thank Mark Fisher of *The Electronic Sky* for supplying the galaxy drawing background. I am grateful to John Anthony West and to Robert Schoch for their recognition of the importance of my initial study of the Nabta calendar circle, and for their continuing encouragement. I thank Paul A. Rosen, of the Jet Propulsion Laboratory, for forthright discussions.

FOREWORD

by Robert M. Schoch

You hold in your hands (or, in this electronic age, perhaps you are reading it as an e-book) a potent volume that could, or probably should, change the very way that you, that we, think about our forebears in the "prehistoric" past (that is, before about 3500 BC). According to the evidence and analyses presented by Thomas G. Brophy, the ancients (in this case, peoples inhabiting the areas of Nabta and Giza in Egypt) possessed, even by our modern standards, truly amazing and advanced astronomical and cosmological knowledge. I have no doubt that many conventional archaeologists and scientists, as well as committed debunkers (a.k.a. skeptics), will immediately dismiss either some or all of Brophy's conclusions out of hand. Indeed, as described by Brophy in his section on peer review, they already have. I am confident, however, that the material presented herein needs to be considered impartially and meticulously. If it continues to hold up to scrutiny, the implications are staggering (as discussed by John Anthony West further in the Afterword). According to Brophy's most "outrageous" interpretations (I can hear the critics now), a people living over seven thousand years ago may have possessed technical knowledge in astronomy and physics more advanced than our current understandings of the same subjects.

*   *   *   *   *

There are three basic aspects or levels of knowledge that these ancients had, as inferred by Brophy in the present book, that I would like to address briefly. These we can list in order of increasing "difficulty" levels; that is, they become increasingly less credible or believable for the traditional archaeologist.

1) Maps and markers denoting objects, alignments, and events that can be observed in the sky with the unaided (naked) eye.

2) Markers indicating celestial phenomena and events that cannot be observed (apparently) with the unaided eye.

3) Detailed astronomical and cosmological information, such as distances to stars, speeds at which stars are moving away from us, the structure of our galaxy (The Milky Way), and information on the origin of the universe, which we have either only just discovered in modern times, or possibly information (for example, concerning planetary systems around stars) that we do not even have available to us at the moment.

I will comment on each of these three categories in turn.

Maps, markers, and alignments of clearly visible objects

Brophy persuasively makes the case that the ancient Nabta stone circle and interior stones are essentially a very easy to use (user-friendly) star-viewing diagram. So far, so good; it is well accepted that ancient peoples could and did view the stars and take an interest in their positions, using their apparent motions (along, of course, with the movements of the Sun and Moon) to devise calendrical systems and so forth. Things become more interesting when we get to the stars referred to by the Nabta circle and the dates that it could have been used as an actual star-viewing chart. According to Brophy's analysis, three of the stones inside the Nabta circle represent Orion's belt, the same portion of the same constellation that Robert Bauval has suggested the three major pyramids on the Giza Plateau represent. Furthermore, the Nabta circle would have applied to Orion's belt as it appeared on the meridian (an imaginary line in the sky running from the north to the south through the zenith that stars cross nightly as they rise in the east and set in the west) each night at around the time of the summer solstice (when the Sun rises from the eastern horizon farthest north and is highest in the sky at noon in northern latitudes) during the period of 6400 BC to 4900 BC. This is an amazingly early date by conventional standards, beating out the traditional date for Stonehenge by a few thousand years, and furthermore is quite compatible with my own "conservative" dating for the earliest portions of the Great Sphinx located at Giza (I have dated the earliest portion, the core-body, of the Sphinx to circa 5000 BC to 7000 BC). Still, by traditional thinking, the Nabta circle as a star map of Orion's belt during the sixth millennium BC is certainly conceivable and believable. It should also be noted that the archaeologist Fred Wendorf and his team, the original discoverers and describers of the Nabta site, dated construction of the stone circle to at least 4800 BC and associated it with the observation of the meridian and the summer solstice sunrise, so Brophy's analysis here is not that much of a stretch from the conventional archaeological interpretation of the Nabta circle that has been published thus far.

But Brophy does not stop here with his analysis of the Nabta circle. Three more stones, until now totally inscrutable, lie in the center of the circle. These, Brophy demonstrates, also refer to the constellation of Orion. However, they represent not the famous belt, but the head and shoulders of the constellation. And that is not all; these stars would have been viewed in the position shown on the Nabta diagram around 16,500 BC! This is a date of such antiquity that debunkers and hardcore conventional academics will immediately stop reading, but Brophy makes the compelling case that this is in fact what these stones represent. Was the Nabta circle originally built over eighteen thousand years ago? Or is it a rebuilding of an earlier structure? Or was it built to commemorate an earlier time, a time and orientation of the stars passed down by tradition or calculated later? (By analogy, Bauval and Hancock have argued that the present structures at Giza, dating to the third millennium BC, commemorate a time and events thousands of years earlier.)

I can already hear the critics say that the constellations as we know them today are a relatively recent concept, and certainly Orion had not yet been conceived eight thousand years ago, much less eighteen thousand years ago. I would point out that this is only an assertion (or more generously, we might say an hypothesis) on the part of the critics, but also has little bearing on Brophy's analysis. Whether or not Orion was acknowledged as a constellation per se, the stars that we think of as Orion were in the sky and could be observed. I would also note here the work of Frank Edge (see article in *Kadath: Chroniques des Civilisations Disparues* [Spring-Summer 1998] as well as discussion in my book *Voices of the Rocks*, Robert M. Schoch with Robert Aquinas McNally, 1999.). Analyzing the cave paintings at Lascaux, in my opinion Edge has made a compelling case that the constellation of the Bull (Taurus) was recognized in about 15,000 BC and has been passed down to us essentially unchanged over the last seventeen thousand years (and of course the origins of Taurus could be even earlier).

When Thomas Brophy first contacted me (11 November 2001) concerning his research on Nabta he sent me a draft of what is essentially Part I of this book. In his cover email he wrote: "I think you may find interest in the results of the attached presubmission manuscript." Upon reading what he had sent, I immediately saw that his cover comment was an understatement (not that he necessarily meant it that way). The results discussed in the last few paragraphs above are extremely important and significant, but I soon learned there was much more to come.

I next received from Brophy a draft of his paper on the Giza monuments, essentially Part III of this book (actually written by Brophy prior to the rest of the book). In this section of the book Brophy, in part, tests and elaborates upon the work of Robert Bauval, namely that the monuments of the Giza Plateau (including the three classic pyramids attributed to the Old Kingdom pharaohs Khufu, Khafre, and Menkaure) correlate with various stars and other celestial phenomena, including the constellation of Orion. A portion of this work by Brophy, although quite sophisticated and even subtle, still falls within my category of "Maps, markers, and alignments of clearly visible objects."

Bauval has suggested a significant date of 10,500 BC associated with the Giza Plateau (correlating with the positions of the three stars in Orion's belt represented by the three major pyramids on the plateau, as well as other phenomena). Bauval has not claimed an actual date earlier than the third millennium BC for the pyramids of the Giza Plateau, although their positions appear to commemorate an epoch some eight thousand years earlier. Independently, John Anthony West and I have been working on the Giza Plateau. We have come to the conclusion that the Great Sphinx (or at least its oldest portions) predates the Old Kingdom (third millennium BC) by thousands of years, and we are also developing evidence that various pyramids (most notably the second or Khafre pyramid, but possibly also the Great or Khufu Pyramid and the third or Menkaure pyramid) are more recent structures built on top of or in the positions of much earlier structures.

Now Brophy has undertaken the most accurate analysis of the possible correlations of the major Giza monuments with celestial phenomena so far. His findings are remarkable, and in the broadest aspects corroborate the earlier suggestions that there are significant dates associated with the Giza Plateau that predate the Old Kingdom by about eight thousand years or more. As you will read below, Brophy has found significant matches between the monuments of Giza and Orion's belt, in particular at 11,772 BC and 9420 BC, as well as various other alignments. I will not elaborate on this aspect of Brophy's analysis here, for he explains it clearly in the pages that follow. For now let me suggest that although he differs in details from the analysis of Bauval, overall the gist of Brophy's work is to corroborate that very early dates are associated with, or commemorated in, the ground plan of Giza and that stars and constellations are involved. Brophy clearly demonstrates that Orion is associated with the Giza Plateau, and the monuments of Giza are markers of astronomical significance that indicate an underlying sophisticated astronomy in very early times—much to the chagrin of many conventional archaeologists and debunkers. Yet all of the analyses that we have discussed up to now are conceivable, if only remotely, to most "conventional" (I use this word for lack of a better term) academics since they involve nothing more than rather astute and perhaps meticulous and painstaking observations of the skies that could have been carried out by "primitive" people. Such is not the case for the next two categories of knowledge that Brophy infers the ancients possessed.

Non-observable celestial phenomena

The core of Part III of this book, on the Giza monuments, is Brophy's inference that the monuments of the Giza Plateau mark the location of the Galactic Center (center of our galaxy) at its northern astronomical culmination (located on a highest or northernmost point on the meridian) in 10,909 BC. This suggestion is remarkable for at least two reasons: a) 10,909 BC is a very long time ago, but it is compatible with the analyses we have been discussing above, and b) the Galactic Center is not observable with the unaided eye (at least not directly, although perhaps it can be inferred approximately with naked eye observations, as discussed by Brophy in the pages below). And why would anyone care about the Galactic Center thirteen thousand years ago anyway?

Brophy suggests that the Giza monuments, using the Galactic Center as a key point, form a grand zodiac clock. This is not a zodiac clock in terms of the months and modern horoscopes, but a clock of very long time scales and precessional ages.

The night sky of today is not the same as the night sky of two thousand years ago or twelve thousand years ago or eighteen thousand years ago. To put it into crude, non-technical terms, Earth wobbles as it spins on its axis. This wobbling movement is known as precession and affects how celestial events appear as viewed from Earth. For instance, most people know that today the north celestial pole is marked approximately by the star commonly referred to as Polaris (Alpha Ursae Minoris, the star at the tip of the constellation Ursa Minor, or the little dipper), but due to precession the spin axis of Earth does not always point to the same position in the sky, so about 3000 BC the north celestial pole was marked by Alpha Draconis (also known as Thuban), and about 12,000 BC it was marked by Vega (which plays a prominent role in the analysis by Brophy of various megaliths located at Nabta, discussed in Part II of this book).

Not only does the apparent position of the north celestial pole appear to change, but due to precession the positions of all the stars relative to Earth and the Sun also change. Thus at present on the vernal equinox (spring equinox in the northern hemisphere, when the Sun crosses the celestial equator from south to north, also known as the first point of Aries, occurring around March 20-21) the Sun rises against the constellation of Pisces. However, due to precession, the stars against which the Sun rises on the vernal equinox slowly change over time. The Sun will rise against Aquarius in the future, and subsequently it will rise against the other twelve zodiac signs in their order along the ecliptic (the apparent annual path of the Sun among the stars)—Capricornus, Sagittarius, Scorpio, Libra, Virgo, Leo, Cancer, Gemini, Taurus, and Aries (note that this is the reverse of the order that the Sun passes through the signs of the zodiac on a yearly basis)—before repeating the cycle again. The complete precessional cycle as viewed from Earth, sometimes known as a "Great Year" or a "Platonic Year," takes approximately 25,800 years in round numbers (it actually varies over time through the ages), and thus on average the Sun will rise against one of the zodiac constellations for about 2150 years before passing into the next constellation (or, in a more astrological context, the precessional year is often cited as the canonical 25,920-year period, of metaphysical and symbolical importance, with each age lasting 2160 years). Today we are in the zodiac age of Pisces, but three thousand years ago the Sun rose against Aries on the vernal equinox, thus the world was in the zodiac age of Aries. Of course, not all the zodiac constellations are of the same size, so if we actually follow their outlines, some of the zodiac ages will be longer and others shorter. In fact, astrologically all the zodiac ages are considered to be of the same lengths temporally and the constellations are just a way of naming or marking the ages; exactly where each age ends and another begins can be a point of contention. In the not too distant future (within the next couple of centuries, depending on where one draws the boundary between the age of Pisces and Aquarius), we will enter the zodiac age of Aquarius.

The Greek astronomer Hipparchus (second century BC) is often credited with discovering, or at least publicizing, the concept of precession. Precession is a very subtle, long-term phenomenon that according to traditional thinking was not or could not have been known to the ancients before the intelligent classical Greeks. However, there is mounting evidence that precession was not only well known to the very ancient ancients (that is, well before the classical Greeks), but that it was of extreme importance to many ancient cultures as expressed in their surviving mythologies (see the classic work by Giorgio de Santillana and Hertha von Dechend, *Hamlet's Mill*, 1969). Brophy adds to the evidence for an early knowledge of precession with his analysis of the Giza Plateau monuments.

Brophy marshals evidence to support his hypothesis that the Giza monuments served, among other functions, as a zodiac clock that marked the end of the age of Virgo and the beginning of the age of Leo at the time of the northern astronomical culmination of the Galactic Center in circa 10,909 BC. (This, some might argue, fits the motif of the Great Sphinx with a human head representing Virgo and a lion's body representing Leo; however, we don't know what the original head of the Sphinx looked like since the current head is clearly a re-carving [see *Voices of the Rocks*]. Another interpretation is that the Great Sphinx marks the Leo-Aquarius opposition, fusing the two opposites of beastly lion and human water-bearer, matter and life fused with spirit, higher consciousness, and the divine.) But, returning to the issue we raised previously, how did the ancients know where the Galactic Center was located thirteen thousand years ago? Again, to reiterate, the Galactic Center is not an object or phenomenon that can simply be sighted (like the Sun or a star) with the naked eye. Or at least it is not such a phenomenon today. While reading this portion of Brophy's work, the research of Paul LaViolette (see his book *Earth Under Fire*, 1997) immediately came to mind. LaViolette has suggested, based on various lines of evidence including the relative intensity of cosmic rays at Earth's surface over the millennia, that the Galactic core emits bursts ("outbursts") of particles and electromagnetic radiation on a periodic basis (around every 13,000 to 26,000 years). According to LaViolette, a major Galactic core outburst (evidenced by a major cosmic ray event) spanning several thousand years climaxed around 14,200 years ago. So, I wonder, was the Galactic Center clearly visible to the astute observer in 10,909 BC at its northern astronomical culmination? Additionally I would note that the Galactic Anticenter, the antipode (opposite position in the sky) of the Galactic Center, is also, according to Brophy's analysis, marked on the giant sky map formed by the structures on the Giza Plateau. Interestingly, the Galactic Anticenter lies near the constellation of Taurus (in a sense, Orion and Taurus can be seen as pointing to the location of the Galactic Anticenter). And remember that Frank Edge suggests that Taurus has been recognized as a constellation since at least circa 15,000 BC. Both the Galactic Center and the Galactic Anticenter appear to have been of significance to the ancient ancients.

Everything we have discussed so far, from the Nabta circle star diagram in circa 5000 BC to the marking of the location of the Galactic Center in 10,900 BC is, in my opinion, quite remarkable but not totally surprising given my own research on the Great Sphinx and related studies. However, Brophy's analyses go even further.

Detailed astronomical and cosmological information

When I first read a draft of Part II of the present volume, I was flabber-gasted. I did not know what to make of Brophy's analysis, and I still am not quite sure what to make of it. If Brophy is correct, we presently have no way that I am comfortable with to explain how the ancients gained the knowledge they apparently had. Perhaps I must remain uncomfortable.

In Part II Brophy analyzes the positions of a number of megaliths located at Nabta south of the stone circle. These megaliths he interprets as representing the same six stars of the constellation of Orion that are repre-sented by stones set inside the Nabta stone circle (the three stars of Orion's belt and the three stars that make up Orion's head and shoulders). He then goes on to demonstrate that these stones are set to mark the vernal equi-nox heliacal risings (when a star rises with the Sun on the vernal equinox) of the Orion stars in question, along with Vega, during the time period of about 6425 BC to 5400 BC. This is certainly interesting, but it is not inconceivable that these megaliths could have been raised to mark heliacal risings using only naked eye observations and extrapolations combined with relatively primitive (stone age) technology. The results of Brophy's further analysis, however, are mind numbing.

The megaliths are arranged at various distances from a central point, known as "Complex Structure A." As you can read in detail below, Brophy interprets these distances from the central point as recording a) the actual distances of the stars in question from our solar system, and b) the speeds that the stars are moving away from us. The information and interpretations that Brophy extracts from the Nabta megaliths correspond to a high degree with modern knowledge of these parameters for these specific stars. Further, analyzing smaller companion stones associated with the primary "star" megaliths, Brophy suggests that the builders of the Nabta site may have had information about planetary systems or companion stars associated with the six stars in question—information that we do not have today! And then there is the "Cosmological cow stone." Does this stone encode information about the origin of the universe, the age of the solar system or universe, the structure of the galaxy and universe, and / or the fundamental constants of nature? Read Brophy's analysis and see what you think.

Now on the face of it, this may all seem quite outrageous, but in the pages that follow Brophy presents the evidence and his analyses. He develops a serious and scientifically testable explanation for the specific placements of the stones in question. And it should be noted that, as astonishing as Brophy's explanation is for these megaliths south of the Nabta stone circle, whether he is ultimately right or wrong in his interpretation of them does not affect the validity of the rest of his work at Nabta and Giza. It should also be remembered that Brophy's analysis is based only on the material that has been uncovered at Nabta thus far; the site has yet to be fully excavated. Further excavations will help to either corroborate or refute Brophy's ideas. And then there are the mysterious and thus far poorly studied stone sculptures of Nabta. To the uninformed eye, some of the worked stones at Nabta (such as the "cow") may appear to be incredibly crude or very abstract representations of animals or other forms, but following Brophy's hypothesis, what we may be looking at is cosmology in stone.

Confronted with Brophy's analyses and interpretations, conventional archaeologists have three options: 1) Ignore it, 2) Deride and dismiss it without having refuted it, or 3) If they are responsible scientists, refute it and provide an alternative explanation, or acknowledge its validity. If Brophy's work stands up to scrutiny, our concepts of what the ancients knew will be revolutionized.

*   *   *   *   *

As a geologist, I am always interested in the question of how old a site or structure is. Even if celestial events of a certain epoch are being recorded, the structure itself could conceivably be much younger. This is an issue that Brophy addresses in the pages that follow. Here I would like to comment that, based on my best understanding of the Nabta site (and without having yet had the opportunity to visit it and study the structures and their geologic context firsthand), I would bracket the various structures under consideration in this book as having been built during the period of about 20,000 years ago (oldest bedrock structures) to approximately 7,000 years ago (various surface megaliths).

So what does one make of all this? It is up to you, the reader, to judge how much of what follows you can accept. But, Thomas Brophy clearly lays out the evidence and his analyses. The Nabta site deserves further investigation and impartial study, as well as careful preservation for future generations.

Robert M. Schoch
August 2002

Robert M. Schoch, Ph.D., is on the faculty of the College of General Studies at Boston University, and author of *Voices of the Rocks: A Scientist Looks at Catastrophes and Ancient Civilizations.*

INTRODUCTION

On a desolate plain in the Egyptian Sahara desert, about 100 miles west of Aswan, there is a very remote prehistoric archaeological site called Nabta Playa. In 1973, archaeologist Fred Wendorf and his team, navigating by compass in a region with no roads, in the middle of nowhere, stopped to take a water break on their way to another excavation site. They looked down and saw numerous artifacts, potsherds, at their feet. Thus began several seasons of excavations at Nabta Playa. There, according to their archaeological specialty, Wendorf's team searched for Neolithic (Stone Age) remains. In 1990, they were again surprised by the reality of the strange location. Several large rocks lay scattered around the desert. For years they assumed these rocks were chunks of bedrock poking through the sand. Closer inspection revealed that they are actually megalithic (megalith is from the greek roots for large stone) constructions and standing stones that had been placed in the sediments by prehistoric peoples. In 1998 *Nature* magazine published their discovery of what is considered to be the oldest dated stone calendar circle with alignments to summer solstice sunrise and to the north-south direction (*"Megaliths and Neolithic astronomy in southern Egypt"*).

My interest in the archaeoastronomy of Nabta Playa was the result of my study of the astronomy of the Giza plateau. Archaeoastronomy simply means study of the astronomical meanings of ancient monuments. I set out to test Robert Bauval et al.'s idea that the Great Pyramids at Giza model the three stars of Orion's Belt. My plan was to use very accurate regressions of star locations, rather than the roughly approximate star locations used by others. I found that the basic idea of the correlation of the

1

three Great Pyramids to Orion's Belt is clearly correct, and further I discovered important additional astronomical significances in the Giza monuments. This finding was so significant, I thought, that I submitted it to a major journal for consideration. The essence of that paper is covered in Part III of this book. The results of that study led me to conclude there might be an important archaeoastronomical site in far southern Egypt or northern Sudan. A search led me to find the *Nature* magazine paper about the Nabta Playa site, at 22.5 degrees north latitude.

Meanwhile, Wendorf's excavators had uncovered more megalithic constructions at Nabta Playa. Their measurements are published in a collection of academic papers: *Holocene Settlement of the Egyptian Sahara*, by Fred Wendorf, Romuald Schild and Associates (2001). From around 7,000 BC to around 4,000 BC there was much human activity at Nabta Playa and the climate was pleasant, with annual rainy seasons during which the playa would fill with water, and dry seasons when it would drain. Over thousands of years, during the human use of the location, sediments built up in the basin now called Nabta Playa. The playa sediments are generally around 8 to 12 feet deep, from the surface down to bedrock. The megaliths discovered by Wendorf's team are located mostly in the playa basin. Some megaliths are strangely sculpted onto the bedrock, covered by eight to twelve feet of playa sediment. Many megaliths are on the surface of the sediments or embedded into the sediment, and one sculpture was carefully buried midway between the surface and the bedrock.

In their collection of research papers published at the end of 2001 and available in libraries a few months into 2002, the *Holocene Settlement* book, Wendorf et al. made available for the first time detailed measurements of some of the megalithic structures. In that book, years after recording the site, they released the measurements, but the archaeologists remained baffled regarding the true meanings or uses of this prehistoric system of megalithic structures.

My study of these megaliths started with trying to understand the calendar circle at Nabta Playa. An insight as to the calendar circle's meaning and use led to a step-by-step sequence of revelations, each step verified by measure and calculation, that the entire system of megalithic architecture at Nabta Playa is a unified and detailed astrophysical map of truly astonishing accuracy, with no less than staggering implications.

This book reports some of the major discoveries, how I discovered them, how these discoveries can be verified by other researchers, and how to decipher the system of megaliths in a way that will surely lead to more discoveries.

A Note on Peer Review:

The peer-review procedure in modern science is founded on a good principle. New discoveries, according to the theory of peer-review, can be submitted to a journal that will have experts or specialists determine whether the new discovery is sound and soundly presented. If it is, so the process should go, the journal will then print the discovery so other interested people who may not be specialists can access the new discovery knowing that a specialist referee concurred that it is a sound discovery. If the procedure works according to plan, peer-review can be a good and useful process. In practice though the procedure can fail, and when it does fail it can be an impediment to progress.

Some skeptics who do not like the finds presented here will try to discredit them on the basis that they don't appear in the usual peer-reviewed magazines. There are two responses to such skeptics.

First, in this case, I tried to engage the peer review system on two parts of this four Part book. I submitted the essence of Part III about the astronomy of the Giza plateau to a major general science journal, in early October 2001, and received a quick expression of intent not to review it. Due to my experience having published several peer reviewed articles myself, and having many times served as a peer-reviewer, I partially expected such a response on that topic. Archaeoastronomy in general and the study of the Great Pyramid in particular have been notorious lightning rods, seemingly discrediting with ease many accomplished and respected astronomers and scientists.

I was surprised though at the response to my second submission. In late October 2001 I submitted what is essentially Part I of this book, about the Nabta Playa Calendar Circle, to another major general science journal. Even given my familiarity with the sometimes flawed peer-review process, I thoroughly expected that submission to have a high probability of acceptance. My reasons for thinking so were these. This calendar circle study found results that are of general interest on the use of the ancient calendar circle as a star-viewing diagram. The findings also are a major advance to our knowledge of the level of cultural sophistication and development of Neolithic humans. The findings are easily reproduced and tested through astronomical calculations. The findings are of general interest and a major advance, but are not so revolutionary as to be a "lightning rod" like anything about the Great Pyramid seems to be. But that journal also refused to even review the paper. Oddly, they based the decision only on an informal consultation with one researcher. Thus I had no opportunity even to reply to formal criticisms: the review procedure did not work successfully in this case.

Those two parts of this book are interesting and important advances but are not truly revolutionary to major fields of science. Those two studies led to Part II of this book: the Origin Map. Discovered chronologically after the other two parts, the Origin Map is a truly revolutionary find. At the time of discovery of the Origin Map, the pattern of failure of the review process on this particular topic was set. In order to engage the review process at this point, I would have needed to start a sequence of appeals, and of waiting for many levels of editors, referees, publishers and possibly attorneys to respond. That would have taken a lot of time and personal expense, and if such an effort managed to get the process to work at all, it would have delayed release of this discovery probably by years. So in this case the public service aspect of allowing the community to know about this find outweighs personal considerations of trying to engage a review system that seemed to be stuck regarding this particular topic.

The second line of response to our hypothetical skeptic who will claim that these finds are not valid *because* they don't appear in a specific peer-reviewed journal is this. The whole point of modern science is that it is supposed to progress independently of subjective pronouncements by human authority figures. Renes Descartes' great revelation that a testable objective reality (as well as subjective reality) does exist ontologically, is true and is the sound basis of modern science. To argue that a given scientific discovery is or is not valid only *because* a given personality said so, would be more like the pre-scientific, pre-rational method. When medieval pre-scientific scholars had a dispute about how many teeth a horse has, they consulted the pages of the authority Aristotle; they never opened a horse's mouth.

Again, the modern peer review system is based on a good principle; it just seems to have had a glitch on this particular topic. Many important scientific advances were originally published without peer review. These include Isaac Newton's revolutionary *Principia* (1687), Charles Darwin's *On the Origin of Species* (1859), and possibly very recently, if it turns out to be important, Stephen Wolfram's *A New Kind of Science*. The book that contains the basic archaeological measurements of the Nabta Playa site, *Holocene Settlement of the Egyptian Sahara*, is also a non-peer-reviewed collection of articles.

So please look in a horse's mouth when you want to know how many teeth he has. Think for yourself, and test my calculations, when you want to decide on the validity of the finds in this book.

PART I

The Calendar Circle

The Nabta Playa calendar circle discovered by Wendorf's team consists of an outer rim of sandstone slabs with four sets of larger gate stones that form two line of sight "windows" in the calendar circle. Inside the circle are six larger stones. The largest of the slabs are almost three feet long, and the smallest are slightly less than a foot. The circle is ten to eleven feet across. When it was discovered, the calendar circle was in partial ruins but the archaeologists were able to reconstruct the ancient locations of the stone slabs.

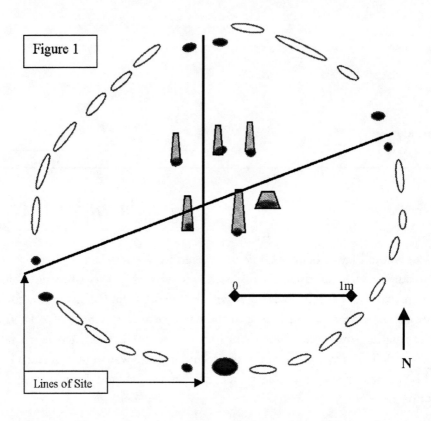

Figure 1

Lines of Site

0   1m

N

Figure 1 is a schematic drawing of the archaeological reconstruction of the Nabta calendar circle. In this drawing, oval shapes are fit to the locations of the calendar circle stones. The outer stones of the calendar circle are shown as unfilled ovals, except for the eight stones that formed two "line of site windows". These eight stones were slightly larger and upright. Also larger and upright were the six stones inside the circle. These central stones are shown in their plan view (looking straight down) by the solid ovals, and the gray shapes show the approximate standing sizes of the central six stones. (This drawing is based on the data in an article by Alex Applegate and Nieves Zedeno, *Site E-92-9: A Possible Late Neolithic Solar Calendar* in the book *Holocene Settlement of the Egyptian Sahara Vol.I,* edited by Fred Wendorf et al.) See the calendar circle photo caption in Appendix C of this book for a description of the determination of the ancient stone locations.

Nature magazine (1998) attributed three significances to the Neolithic stone calendar circle: 1) a meridian sight line window (north-south); 2) a sight line window consistent with viewing the summer solstice sunrise (east by northeast); 3) location of the circle at 22.5°N latitude such that the sun passed through the zenith (straight up) about three weeks before and after summer solstice. And a construction date of at least 6,800 years ago was attributed to it.

In the center of the calendar circle were placed two sets of three large stones. The significance of these large stones has remained unknown. After a few days of puzzling over possible meanings of the central stones of the calendar circle, a sudden insight led me to a series of calculations that clearly deciphered it. Following is what I found to be the significance of those six central stones.

STAR VIEWING DIAGRAM

Three of those central six stones are a diagram of the constellation of Orion's Belt as it appeared on the meridian around summer solstice from 6,400 BC to 4,900 BC. The "meridian" is simply the line in the sky that passes north to south, and is thus the midway line or "meridian" across which stars pass in their nightly travels from rising in the east to setting in the west. Thus the inside of the calendar circle had a clear use as a star-viewing diagram. A user of the calendar circle diagram, standing at the north end of the meridian sightline window, would look down on the stone diagram and see a representation of the stars of Orion's Belt just as they appear on the meridian in the sky when the user looked up.

More than just matching the appearance of the stars of Orion's Belt at that time, the ground stones indicate the sky stars at the special time of summer solstice as identified by the solstice sightline window in the calendar circle.

The stars of Orion's Belt fade out of visibility just before sunrise on the meridian as indicated in the calendar circle diagram, on the days around summer solstice when the sun would then rise in the sightline window of the calendar circle.

Around 4,900 BC the Belt stars passed achronal culmination, meaning they appeared in the sky on the meridian after sunset, out of the dusk twilight, on winter solstice. This indicated the end of applicability of the calendar circle star diagram because the Orion's Belt stars would no longer be visible in the sky on the days and times indicated by the solstice and meridian sightline windows.

Around 6,400 BC Orion's Belt transited the meridian as shown by the calendar circle diagram around 50 minutes before the summer solstice sunrise, fading into the predawn twilight. This is the first and only time that visual fade out of Orion's Belt occurred on the meridian actually on the day of summer solstice as seen at Nabta Playa. (The sky begins to brighten more than an hour before sunrise, and becomes too bright to see most stars around 40 to 50 minutes before actual rise of the sun disk.) So roughly 6,400 BC marked the beginning of applicability of the calendar circle star-viewing diagram for Orion's Belt.

So the northern three central stones make a star-viewing diagram for Orion's Belt around 4,900 to 6,400 BC. The apparent significance of the other three central stones in the calendar circle is more difficult to reconcile with prevalent assumptions about prehistory. These other three stones are a diagram of the configuration of Orion's head and shoulders as they appeared on the meridian on summer solstice sunset in the centuries around 16,500 BC. That date is symmetrically opposite the 5,000 BC congruence of the Orion's Belt stars, in terms of the precession of the equinoxes, and both dates are at the extremes (maximum and minimum) of the tilt angle of the Orion constellation. Thus the stone diagram illustrates the time, location, and tilting behavior of the constellation of Orion through the 25,900 year equinox precession cycle, and how to understand the pattern visually.

Calendar circle stellar astronomy

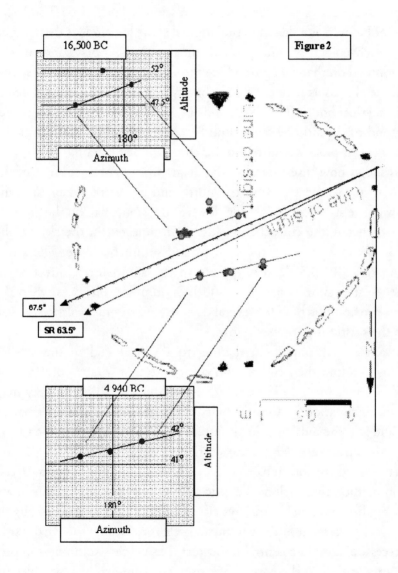

Figure 2 shows the calendar circle stone diagram with the configuration (angle and aspect) of the constellation of Orion's Belt as it appeared on the meridian around sunrise near summer solstice in 4,940 BC. The background of Figure 2 is a sketch of the original stone locations as determined by the archaeologists, and printed in the *Nature* magazine article. Superimposed on that background are the Orion's Belt and Orion's shoulder and head stars as they appear in the sky with the star locations determined in this book. The N-S and E-W coordinates of the stones correspond to the altitude and azimuth coordinates of the stars, only the scale of the star plots is adjusted to match the stone diagram.

This shows how "user friendly" the diagram was. A human user of the stone diagram would stand on the north end of the meridian sightline window of the stone circle, looking down at the stone map on the ground. He/she then saw the constellation just as it appeared in the sky on the meridian before sunrise on the days around summer solstice. The arrow pointing to "SR 63.5 degrees" points to the approximate direction of summer solstice sunrise around 4,900 BC. Summer solstice is when the sun rises furthest north, so for several days around summer solstice the sun rose in the solstice sightline window.

The ancient diagram user, moving to the west end of the solstice observing window, then saw the sun rising into that window on the days around summer solstice. The date 4,940 BC matches the stone sky map when the inclination of Orion's Belt on the meridian is at its smallest angle, and it is the only time in about 25,900 years that Orion's Belt is on the meridian near summer solstice sunrise.

I determined the locations of the stars using a long-term accurate calculation method that includes ecliptic and celestial pole motions, following the method of French astrophysicist Berger (1976, 1978). The roughly 25,900 year precession of the equinoxes cycle is caused by the Earth's spin axis precessing around a central point just like a child's spinning top precesses. In addition to the basic precession, the spin axis also wobbles up and down, changing Earth's "obliquity"; and the Earth's orbital plane about the sun also wobbles, changing Earth's "ecliptic pole". All these motions affect the apparent locations of stars in the sky. The actual star locations and more details on the calculations are given in the Appendices to this book.

CORROBORATION OF THE STAR VIEWING DIAGRAM

The certainty of this interpretation of the calendar circle diagram depends largely on corroborating evidence. The star viewing diagram meanings are magnificently corroborated by the rest of the Nabta Playa site, discussed later in this book. But for now I describe how the star-viewing diagram is corroborated by the calendar circle itself.

There are two sightline windows in the stone calendar circle: solstice and meridian. The matching of the diagram with Orion's Belt coincides with both of the sightline windows. Also, that this occurs at an extreme value of the angular orientation of the constellation further evidences that the asterisms (Orion's Belt, and Orion's shoulders and head) are marked at distinct and unique times, forming a pattern of design.

Thus at Orion's Belt's smallest inclination date in the precession cycle, on 4,940 BC, it was also at vernal equinox in the sky, so that it transited the meridian 41 minutes after sunrise on summer solstice at the latitude of Nabta, rather than exactly at sunrise as it would at the equator. Ten days after summer solstice it was on the meridian at sunrise. 21 days after summer solstice, on the day when the sun transited the zenith (straight up), it was on the meridian 52 minutes before sunrise—fading into the dawn twilight before the sun rose through the northeast sight line window. In following days of the year it was on the meridian further and further behind the sunrise, until winter solstice when it appeared out of dusk twilight on the meridian 35 minutes after the sun set into the southwest end of the solstice window. Then it was no longer visible on the meridian until the next sun zenith-passing date 23 days after summer solstice.

A similar annual pattern was visible for several centuries before 4,940 BC. Back to about 6,400 BC Orion's Belt transited the meridian as shown by the calendar circle diagram around 50 minutes before the summer solstice sunrise, fading into the predawn twilight. Thus heliacal culmination at summer solstice in 6,400 BC heralded the start of the applicability of the calendar circle diagram. ("Heliacal" means first appearance of the year before sunrise. "Culmination" is when a star passes the meridian because it is at it's highest altitude at that time.) And achronal culmination at winter solstice in 4,900 BC indicated the ending date of the calendar circle diagram. ("Achronal" means last appearance of the year after sunset.)

The meridian and solstice sight line windows told the diagram user when and where in the sky to look, and the central stones showed just what to look for. In modern industrial terminology, the diagram was "ergonomically designed" so the user could easily see how to use it even without an instruction manual. If an astute ancient sky watcher from some foreign culture stumbled upon the Nabta calendar circle, even if he had no concept of the constellation of Orion, he would very likely have figured out the meaning of the diagram, as long as he was there during the diagram's window of functionality from 6,400 BC to 4,900 BC.

After our Neolithic skywatcher/archaeoastronomer easily deciphered the Orion's Belt match, he would continue to decipher the diagram. The meaning of the southern three stones in the diagram remains to be discovered. It is immediately tempting to associate them with the head and shoulders of Orion because those stars have configuration similar to the diagram stones. There are no other matching sets of three bright stars near the celestial coordinates in question. But around 5,000 BC, our sky watcher would note, the angle of Orion's head and shoulders in the sky is tilted opposite from the orientation of the stone map. However, precessed to their lowest sky angle these stars match the stone map, as is shown in Figure 2. (Don't say the ancient sky watcher couldn't have known that – I'm getting to that later.)

16,500 BC, 30 minutes before winter solstice sunrise Orion's head and shoulders (the stars Betelgeuse, Bellatrix, and the unresolved group of stars Meissa) were on the meridian with the configuration shown in the calendar circle diagram, disappearing into the dawn twilight. It was visible on meridian transit the rest of that year until 15 days before summer solstice, on the day when the sun transited the zenith, when it appeared on the meridian out of the dusk twilight 25 minutes after sunset. Thus with winter solstice heliacal culmination and summer solstice achronal culmination, the head and shoulders diagram in the stones is the converse of the Orion's Belt diagram, and both are congruent with the solstice and meridian sightlines in the stone calendar circle. As with the Belt stars reaching their angular minimum when their sky precession position was at vernal equinox in 4,940 BC, the Shoulders angle reached angular maximum when their sky position was at autumnal equinox in 16,500 BC, matching the stone map. Also the very bright star Betelgeuse, far brightest of the six stars in question, is represented by the largest stone in the diagram.

Thus around 16,500 BC the diagram was equally ergonomically designed for use by a sky watcher, and it's meaning was a mirror image of the 5,000 BC meaning. But, how could our 5,000 BC Neolithic wandering astronomer have possibly figured out this half of the diagram without a computer to precess the stars back to 16,500 BC? I will argue later that there is a way he could, but for now this is yet another internal corroboration of the star viewing diagram.

Outer circle of the diagram

At first glance, the outer ring of stones of the calendar circle seems crudely planned—not very round. The whole east half of the circle however is rather precisely circular. The west half of the ring, surprisingly, seems adjusted to scale to fit the Orion congruencies. Figure 3 shows the circle layout with the whole Orion constellation as it was on the meridian on the two matching dates. The scale of the circle and the scale of the two matching constellations all relate so that on both dates the boundary stars of Orion fit the deformed circle. On 16,500 BC both of Orion's feet are on the circle with left foot on the edge of the meridian window, and his up-stretched hand is on the stone circle as well, as in Figure 3b.

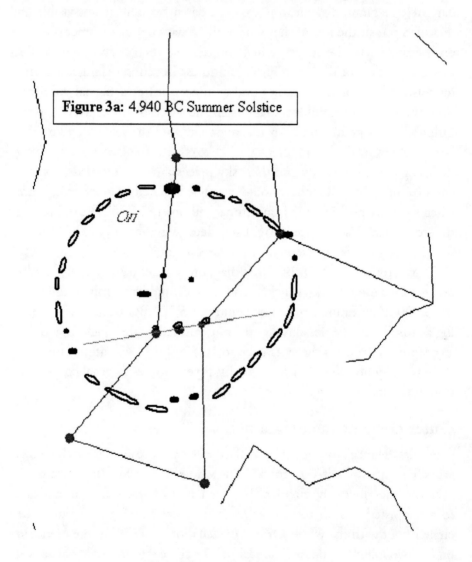

Figure 3a: 4,940 BC Summer Solstice

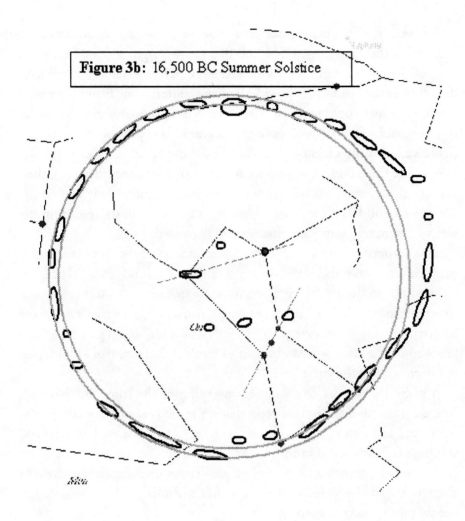

Figure 3b: 16,500 BC Summer Solstice

An overlay of two exact circles in Figure 3b shows the circularity of the east half of the stone ring. On 4,940 BC, Orion's right shoulder is on the edge of the solstice sightline window (Figure 3a). It would be possible to design an exactly round circle with the same central stone fittings but such a configuration would make the two central sets of three stones mix closer together, and therefore be a less clear diagram. Also possibly the different figure scales of the Orion figure might have meaning. On the 4,940 BC match date Orion is half way up its ascent in the sky to northern culmination: thus it is matched with ground figure size conceptually increasing. On the 16,500 BC match date Orion is half way down its descent in the sky, with ground figure size conceptually decreased.

Yet another corroboration of this interpretation of the stone diagram is the location within the circle of each set of three central stones. The locations of the stones in the circle correspond to their stars' respective altitudes in the sky on their respective diagram-matching dates. This is if the length of the sightline corresponds to 90° of altitude, and if ground level is taken to be the place where a human user of the diagram would stand, on the north end of the circle.

A possible, but less corroborated, meaning of the stone circle is this. The diagram marks the same constellation at two extremes: when it is at the vernal and autumnal equinoxes, both half way between northern and southern culmination. On the precisely circular half of the circle, there are twelve stones (counting the sight line window as one) suggesting that the diagram might also signify the partition of the Zodiac into twelve ages and the year into twelve months.

Cultural continuity for the calendar circle

This interpretation for the Nabta calendar circle fits with other correspondences found between the Nabta-Egyptian Neolithic culture and the later Old Kingdom Egyptian cultures occupying essentially overlapping territories. Wendorf and Schild (1998) note that several elements of Old Kingdom Egyptian culture may have come from the Nabta culture, including "the role of cattle to express differences of wealth, power and authority, the emphasis on cattle in religious beliefs, and the use of astronomical knowledge and devices to predict solar events". Dynastic Egyptians used "decans-stars" and star groups to measure time in one hour intervals as they passed the meridian (Trimble, 1964), and they depicted such star groups in art and monuments and aligned a major ritual shaft in the Great Pyramid to Orion's Belt (Badaway, and Trimble, 1964). Dynastic Egyptians signified great importance to the constellation Orion, calling it *Sahu* and associating it with the primary deity Osiris (Mercer 1946). These findings about the Nabta calendar circle further solidify the case for continuity of cultural influence from Neolithic Nabta Egypt to Old Kingdom Egypt which is more that 2,000 years younger.

CALENDAR CIRCLE DISCUSSION

The case for the three northern stones in the diagram representing Orion's Belt on the meridian around solstice from 6,400 BC to 4,900 BC is extremely strong. The case for the southern three stones in the diagram is also strong, but it is troubling to some investigators because there is no evidence yet of activity circa 16,000 BC at Nabta Playa, and the calendar circle couldn't be that old because it sets on younger sediments, and there is bias against the idea that ancient people could have known that the constellations change tilt long term due to precession of the equinoxes. The last point is certainly unreasonable though. Given that the Late Neolithic people also built precise long distance megalith alignments to rising stars, it is very likely they knew the stars changed rising locations over time. To the precision that they were aware of, less than a degree, and perhaps even to a few minutes of arc, given that they seem to have marked east to within 0.02°(ref. *Nature* 1998) the stellar precession motion could have been apparent even within a human lifetime.

And because they marked northerly and southerly stellar risings, they could easily have deduced that the whole constellation of stars tilted as well as changed rising azimuths, without using any mathematics, even if they were as dumb as I am. For them to have deduced the maximum of tilting (as is represented in the diagram) would have been a bit more difficult but not impossible. If, as Wendorf and others argue, they assigned importance to the solar zenith crossing (the two days a year when the sun passed directly over head) and tracked the sunrise azimuth throughout the year as indicated in the calendar circle, they had the elements in place to deduce from conceptual thinking alone the long term tilting of Orion, which is about 55° basically the same as the azimuth shift (the angle measured on the ground) from summer solstice to winter solstice sunrise. Another possible scenario is that much more ancient observations of constellations were passed along culturally. Thus Late Neolithic peoples could have designed the calendar circle around 5,000 BC to mean what it appears to mean, even with only the wherewithal that scholars generally ascribe to that time: no mathematics and little technology but modern brains and careful observation.

Finally another strange, perhaps coincident, aspect of the calendar circle is that around the previous time when Orion's Belt was at descending equinox, earth's axis was near a minimum obliquity very close to 22.5° which is the latitude of Nabta Playa. In fact, running Berger's accurate celestial pole model to the year when the Earth obliquity exactly equaled the latitude (31,330 BC), it was the time when Orion's Belt again matched the stone diagram, congruent on the meridian 15 minutes before summer solstice sunrise. One reviewer of my initial submission of this study said the fact that I even think about these old dates weakens my case for the 6,000 to 5,000 BC diagram interpretation. It should not. I am not claiming that the current calendar circle is that old. All interesting aspects of a study such as this should be reported and not overly self-censored. This strange coincidence of the diagram repeating and being even better at an older time exists, and might possibly add value to understanding the site as more data is gathered. Simply mentioning such facts should not weaken the very strong case for the 6,400 BC to 4,900 BC congruencies.

Other evidence that may possibly suggest very ancient activity at the site was reported by Wendorf and Schild (1998). The Late Neolithic inhabitants at Nabta Playa were able to locate and excavate subterranean "table rocks", several meters beneath the surface. These table rocks would have only been apparent several thousand years earlier before being covered by playa sediments, unless those subsurface rocks were somehow marked previously, as they were marked in the late Neolithic for Wendorf and team to discover them now in the post-modern.

Finally, as the next part of this book will show extensively and clearly, this calendar circle interpretation is magnificently corroborated by the rest of the megalithic complex at Nabta Playa.

The design of this star diagram in the calendar circle stones is wonderfully simple and clear. Whether this really is the earliest astronomical stone circle or not, it must rank among the most elegantly and artfully designed.

PART II

The Origin Map

The calendar circle analysis, with the technical details in the appendices, is what I submitted to a major journal for peer review. The journal refused to review the paper, on advice of one researcher whom the journal editor consulted informally. While I was dumbly mulling over that strange sort of nonreview-review-rejection all in one, I was considering where else I might submit the calendar circle study. A few months later the book (*Holocene Settlement of the Egyptian Sahara*) containing a paper with data on the long distance megalith alignments at Nabta Playa became available. In that paper was printed the actual GPS location measures for some of the megaliths. And for the first time they published the locations of two additional sets of megalith alignments in the Nabta Playa complex. Other papers by Wendorf et al. in the *Holocene Settlement* book present the basic archaeological data on several other important features at Nabta Playa. Following is how I found that the megalithic complex at Nabta Playa perfectly corroborates my interpretation of the calendar circle as a star-viewing diagram.

MEGALITH ALIGNMENTS

Near the calendar circle, the Neolithic builders of the Nabta Playa site

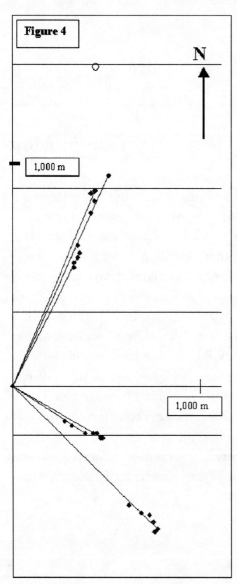

Figure 4

N

1,000 m

1,000 m

placed several large stone megalithic slabs that range in size from a few feet in length to about 12 feet long, with the largest weighing about 10 tones.

The megaliths were lying on or embedded in the playa sediments. They were quarried someplace, brought to the location, embedded in the sediments, and most of them were also shaped. Most of the megaliths are now broken, either intentionally or by natural causes. Wendorf's team determined their original dispositions and measured their locations. The designers of the system of megaliths placed them in straight lines radiating outward from a central point marked with more megaliths. There are many megaliths in the Nabta Playa basin. So far, Wendorf et al. have published the precise locations of 23 of them.

The locations and alignments of those 23 megaliths are plotted in Figure 4, and given in the table in Appendix B. These are precise locations measured by Global Positioning Satellite (GPS) system measurements.

The circle to the north in Figure 4 represents the calendar circle. The calendar circle location is only approximate, as measured from the plots in the *Nature* article that is known to contain errors. The megalith locations however, from the GPS data in the more recent Wendorf et al. book, are very accurate. The six lines represent those sets of megaliths identified by Wendorf et al. as clearly aligned together. The names assigned to the megaliths are given in the table in Appendix B, with their GPS location data.

Megaliths mark the calendar circle stars

When I received the GPS location data for the megaliths (May 23, 2002), and the revelation of the two additional sets of megaliths (B1, B2) that were not published in the *Nature* paper, it was immediately apparent that these alignments might corroborate the calendar circle meaning. Even cursory initial consideration of the locations of the six Orion stars in question (Alnitak, Alnilam, Mintaka, Betelgeuse, Bellatrix, Meissa) suggests that the megaliths probably represented them.

The rising directions of the six Orion stars represented in the calendar circle diagram are indicated by the southern long lines in Figure 5. Vega, the brightest star in the north, is indicated by the northern long line. Two of the northern megaliths and one southern megalith are indicated with hollow, rather than solid, symbols because their separate meanings will become apparent.

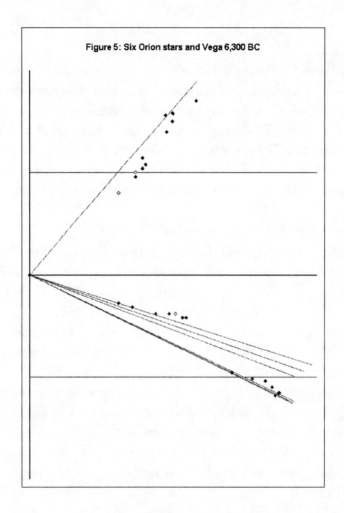

Figure 5 shows the rising directions of all six Orion stars on the same date, 6,300 BC. One can see that all six stars represented in the calendar circle were close to one or more of the megalith alignments. This means that, within a few hundred years of each other, some time around 6,000 BC all six of the stars aligned with one or more of the megalith lines. And simultaneously the brightest star in the north, Vega, rose with the northerly megalith alignments.

When I first looked at this pattern I was excited to see that the megalith alignments might point to the same stars as are indicated in the calendar circle stone diagram, and they also aligned around the same dates as the calendar circle star-viewing diagram (6,400 to 4,900 BC). But it seemed disappointing that the stars didn't all align at exactly the same time. If they had aligned simultaneously I thought that would have absolutely proven my case for the calendar circle meaning. However, closer study reveals that the corroboration of the calendar circle meaning is even more precise and extensive than a simple simultaneous alignment would be.

Sophisticated star chart

Because I had created a program that calculates the location of Earth's celestial pole very accurately through time, and also gives the locations of stars in ancient times, I was able to consider these alignments in more detail.

Astronomers use two coordinates to locate stars in the sky: *declination* is angle towards north from the equator, and *right ascension* is the hour angle from the sun on the day of vernal equinox. A look at the right ascensions of the Orion stars around 6,300 BC revealed that they were near vernal equinox heliacal rising as seen from Nabta.

Heliacal rising is when a star rises at the same time as the sun rises. Vernal equinox heliacal rising is when this occurs on the first day of spring, the vernal equinox. In fact, all six of the stars were near their vernal equinox heliacal risings, and Vega was near autumnal equinox (first day of Fall) heliacal rising. This suggested a possible more specific pattern in the megalithic alignments, more than just star rising markers.

Vernal equinox heliacal rising of a given star, at a given location on earth, occurs only once per equinox precession cycle, about 25,900 years. So if megalithic structures are oriented specifically to vernal equinox heliacal risings, rather than simply rising directions, the alignments are very specific, and the statistical significance of the alignments is vastly greater. (The chance is small that such alignments occur unintentionally, by happenstance.)

Looking at the exact dates of vernal equinox heliacal rising of each of the six stars represented in the calendar circle reveals the first truly astonishing aspect of Nabta Playa. The following seven illustrations show the vernal equinox heliacal rising alignments of the seven stars.

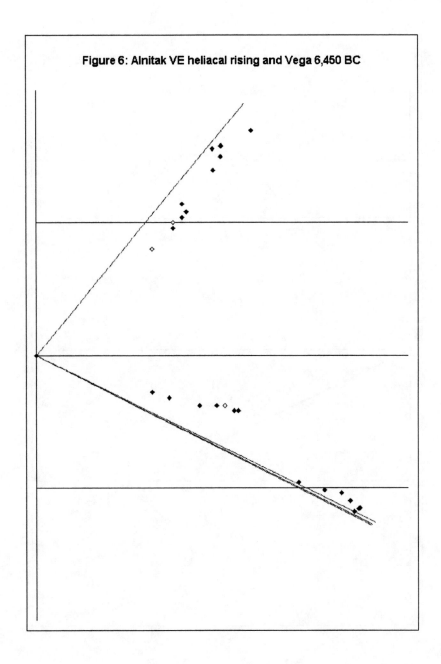

Figure 6: Alnitak VE heliacal rising and Vega 6,450 BC

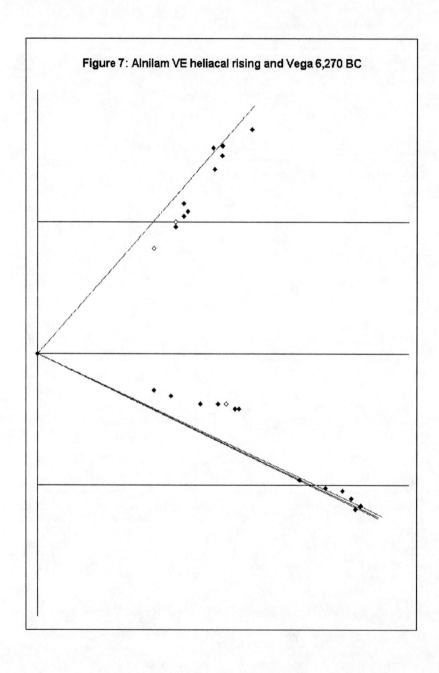

Figure 7: Alnilam VE heliacal rising and Vega 6,270 BC

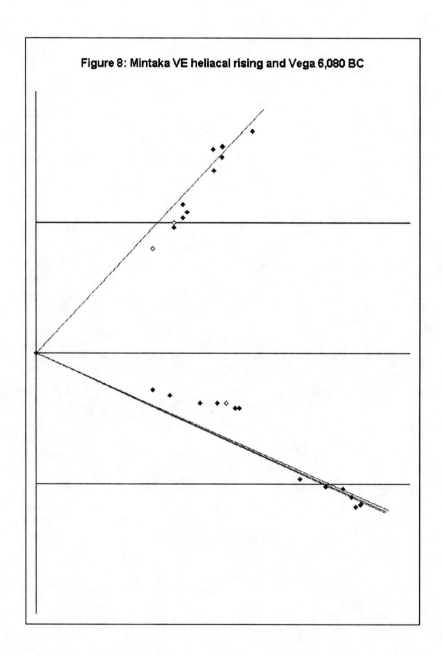

Figure 8: Mintaka VE heliacal rising and Vega 6,080 BC

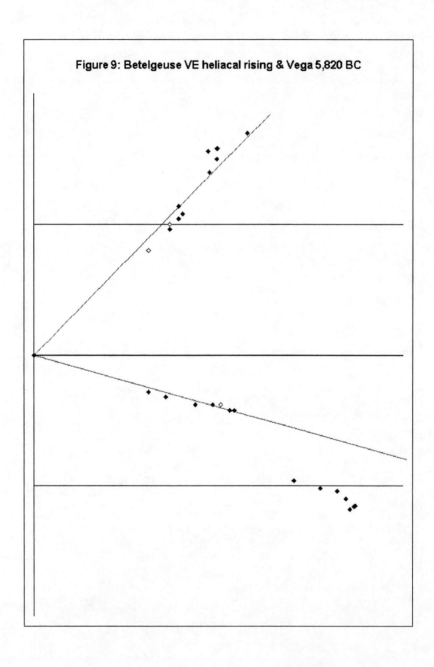

Figure 9: Betelgeuse VE heliacal rising & Vega 5,820 BC

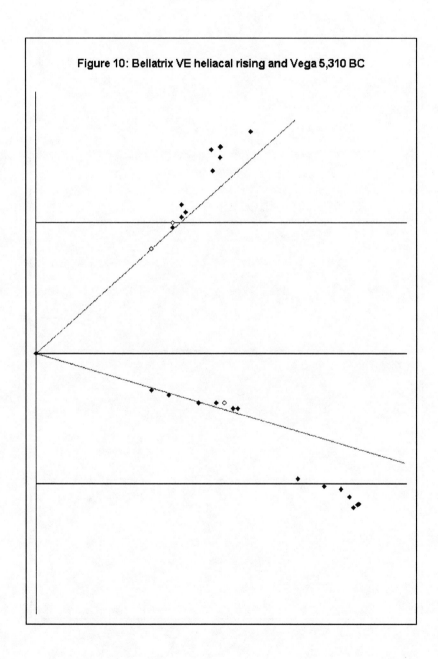

Figure 10: Bellatrix VE heliacal rising and Vega 5,310 BC

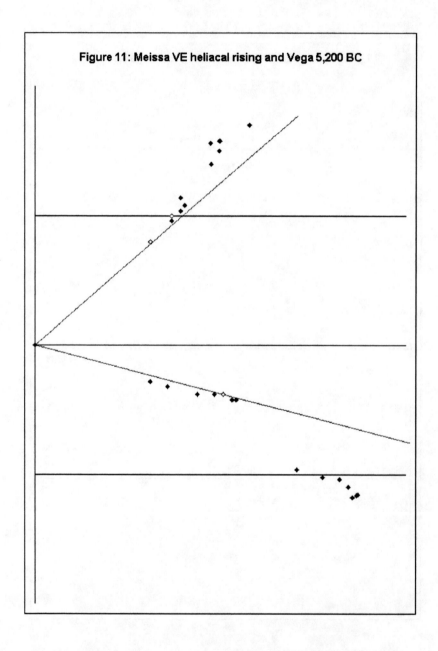

Figure 11: Meissa VE heliacal rising and Vega 5,200 BC

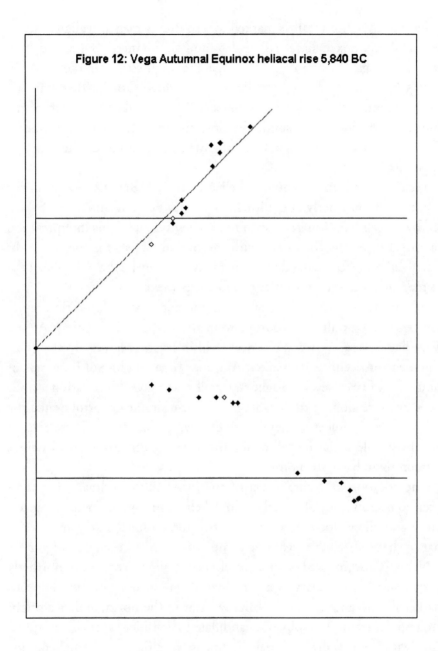

Figure 12: Vega Autumnal Equinox heliacal rise 5,840 BC

Each of the Orion stars is marked by a southerly megalith alignment on its vernal equinox heliacal rising date and simultaneously marked by a northerly megalith alignment with Vega, the brightest star in the north. In figures 6, 7 and 8 rising sight lines to the three Orion's Belt stars are shown together on the vernal equinox heliacal rise dates for each of the three stars: Alnitak is the southerly most line and Mintaka the northerly. The stars corresponding to each alignment are listed in the table in Appendix B.

Vega's own autumnal equinox heliacal rising, Figure 12, occurs in the midst of the northerly megalith lines and is possibly indicated by the nearby megalith A-7 marked with the hollow symbol. This interpretation for megalith A-7 is consistent with the overall system for the megaliths described in the following development, as A-7 may be the only one of the 24 megaliths that doesn't otherwise fit the pattern.

In general, just a simple megalith alignment at a given location represents essentially only one coordinate in the sky, the declination. At any given time several stars will rise near (within several degrees of) any given alignment or declination marker. And in a time window of hundreds or thousands of years several bright stars will pass a given declination as they move by precession. Further calling into doubt simple uncorroborated rise line markers are uncertainties due to observing conditions. These reasons are why simple stellar megalith alignments have been considered dubious in rigorous archaeoastronomy.

The addition at Nabta Playa of the other sky coordinate, the right ascension, specifically at vernal equinox heliacal rising, removes the uncertainty and allows for a very accurate two dimensional star map. Clearly that is what the Nabta megalithic complex is. Further, each star alignment at Nabta is also marked with a simultaneous alignment to the northerly marker star Vega. Vega was a very logical star to use as a simultaneous marker star because it was the brightest star in the north, and as a bright circumpolar star it is an obvious candidate to consider for such a purpose. Thus with the two dimensional alignments specifically to vernal equinox heliacal rising, and simultaneous Vega markings, and the corroboration with the calendar circle diagram, the possibility that these alignments are random chance is essentially eliminated. They probably are the designed meaning of the megalithic complex.

Heliacal events and atmospheric extinction

Heliacal astronomical events are events that occur with the rising sun (named after the Greek sun god Helios). In archaeoastronomy, heliacal rise alignments are generally considered to have large uncertainty because viewing conditions change and there is a broad range of possible definitions of twilight, when the sky becomes too bright to see stars. (Schaefer 1986, 1993). But the heliacal rise alignments plotted here are for geometrical rise with the sun rather than visual sighting before sunrise, thus eliminating the large uncertainty introduced by definitions of twilight. The small corrections for atmospheric extinction, included here, follow the method described by Schaefer to correct for the extinction of a star before it actually sets below, or rises above, the horizon. I use an extinction value of $k=0.25$ which is considered reasonable for Nabta Playa at the time. Nevertheless several factors combine to make the extinction correction for Nabta fairly small: the stars (except for Meissa) are bright; the latitude of Nabta is low so the rising angle is not too far from vertical; and the extinction is moderate. The resulting extinction corrections in azimuth generally range from 1.0 degrees for bright Betelgeuse to about 1.6 degrees for Bellatrix with Orion's Belt corrected about 1.3 degrees, and only dim Meissa significantly corrects by almost 4 degrees.

The visual magnitudes (inversely related to star brightness) for Vega, Betelgeuse, Bellatrix and Meissa are taken to be (0.03, 0.58, 1.64, and 3.3) respectively. For visual extinction correction purposes, the visual magnitudes of the Orion's Belt stars are combined from their separate magnitudes averaging about 1.8 to a combined magnitude of about 0.8, three times as bright.

The extinction corrections occur systematically all in the same direction. So variability in the atmospheric extinction of the stars could alter the alignment dates by several tens of years, but would maintain the overall alignment patterns.

Random chance probability

The probability that these stars aligned with the megaliths by random chance can be estimated according to the method developed by Schaefer (1986). Details of this calculation are given at the end of Appendix A. Many estimates are involved in such a calculation. The more conservative of the range of estimates gives a random chance probability of these seven stars aligning with the megaliths according to this system of less than 2 chances in a million. That is more than a thousand times as certain as the usual 3 standard deviations requirement for accepting a scientific hypothesis as valid. The more liberal of the range of estimates gives a random chance probability of about one in ten to the thirteenth power, or about as likely as picking at random the same human being out of all people on earth twice in a row. By even the more conservative estimate, these are by far the most certain ancient megalithic astronomical alignments of any yet known in the world.

How could they have done this?

The heliacal risings aligned with the megaliths are geometrical heliacal risings on vernal equinox. That is when the stars rise above the horizon physically with the sun on the first day of spring. Visual heliacal rising is when a star first becomes visible above the horizon just before the predawn sky becomes too bright to see it. Visual heliacal risings are thus far less precisely determined than are geometric heliacal risings. Geometrical heliacal risings are not strictly visible because the sun is too bright of course.

However it should not be assumed that ancient sky watchers could not determine geometrical heliacal risings, even without the use of technology or mathematics, if they wanted to. An astute observer would simply watch the rising of the star for several days during the year while the star's rise was clearly visible. If the ancient sky watcher knew when vernal equinox was, he would simply extrapolate the star rise time of day to vernal equinox and determine if it would be a geometrical heliacal rising. That is a very achievable feat for an accomplished ancient megalithic astronomer even without technology or mathematics.

Thus so far in this analysis of the Nabta megalithic astronomy these alignments could possibly have been designed by ancient peoples with the usually assumed low level of Stone Age technology. The designers displayed astonishing elegance and clarity, but plausibly they could have done it with primitive technology.

Three dimensional star map

Given the accuracy and elegance with which the Nabta megalith builders created their star chart, I realized there might still be more information to be deciphered in the chart. As can be seen in the figures, the distances of the megaliths in each line are not placed at a uniform distance away from the central point. If the varied distances didn't have a purpose, one would expect the skilled Nabta Playa designers to have used a more pleasing arrangement. And the distance placements are suggestive of a meaningful pattern.

I first considered whether the visual magnitude of the stars (their apparent brightnesses in the sky) might correlate to the megalith distance pattern. While Betelgeuse is much brighter than the Orion's Belt stars, it is also much brighter than Bellatrix, and Bellatrix is about the same brightness as Alnilam. So visual brightness doesn't explain the megalith distances.

Pursuing further what one would expect to be the logical meaning of the alignment distances in a chart, namely the *actual distances* to the stars, just for fun I looked up our current measures for the astrophysical distances to these stars. The best star distances astronomers have been able to measure are from the Hipparcos Space Astronomy satellite parallax measurements. Astonishingly, these Hipparcos astrophysical distance measurements match the megalith distance pattern quite well.

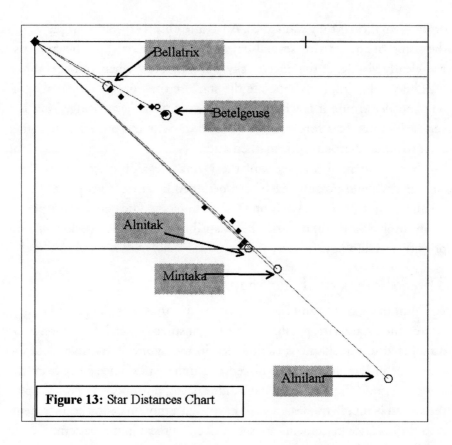

Figure 13: Star Distances Chart

In Figure 13 the distances to Betelgeuse, Bellatrix, Alnitak, Alnilam and Mintaka are shown with empty circles, plotted over the megalith locations. The scale is one meter on the ground at Nabta equals 0.799 light-years in the sky (a light-year is the distance light travels in one year). The correspondence of Betelgeuse and Bellatrix to a megalith in their respective alignments is very good.

Meissa is not plotted. Meissa's parallax-measured distance in the Hipparcos tables would put it much farther than the Meissa megalith, about to the distance of Mintaka. However, the head of Orion and not necessarily the star we call Meissa is what the megalith alignments signify, and Orion's head is actually multiple stars including Meissa. Meissa and its companion star have equal brightness, making their parallax distance measurements more uncertain, and at least one more star visually unresolved from the others with unknown distance makes up Orion's head. Also in the visually unresolved region of Orion's head, clusters of dim low mass young stars are present. One or more of these components could be in the distance range signified by the diagram, especially possibly the Lambda Orion star-forming cluster.

Alnilam appears too far for the diagram, but the Hipparcos measures for Alnilam are the most uncertain so the actual distance to Alnilam is someplace within a large range around the value shown in Figure 13. If we replot the distances with Alnilam, Alnitak and Mintaka at one standard error from their measured values, and use the errors quoted by Hipparcos, all five stars are seen to be within a standard error together of their corresponding megaliths.

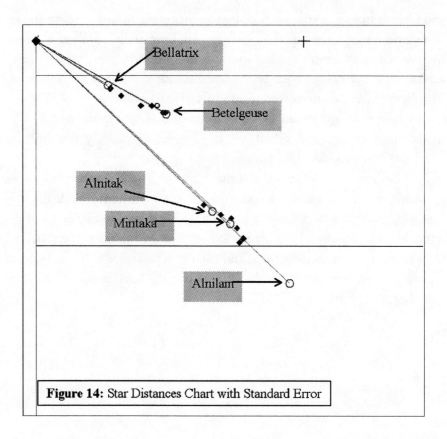

Figure 14: Star Distances Chart with Standard Error

This is more than astonishing. This is the first very important result of this analysis of the Nabta Playa megalith map. Star distances are difficult to measure. Modern science's best estimates of star distances, based on astrophysical models of star evolution, were very erroneous until recently. Only with the launching of the Hipparcos satellite observatory, above the atmosphere, have we been able to directly measure parallaxes (parallaxes are the different angles measured to a star as the Earth travels around the Sun) and achieve somewhat accurate star distances. If these star distances are the intended meaning of the Nabta Playa map, and are not coincidence, then much of what we think we know about prehistoric human civilizations must be revisited. Further study of the Nabta Playa megalith map proves that this is in fact not coincidence.

Star velocities map

Given that the southern megalith placements represent distances to the stars, it is likely the northern megalith placements also represent something. If one thinks like an astrophysicist, and remembers that the northern alignments are time markers for the southern direction alignments that point directly to the respective stars, one is drawn to consider whether the northern placements might represent radial velocities. Radial velocities are the speeds with which the stars are moving away from us.

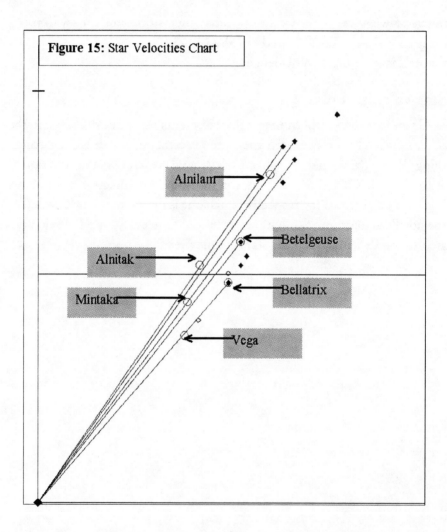

Figure 15 shows the star velocities; from the Hipparcos satellite observatory data, plotted with empty circles. The scale is such that one meter on the ground in Nabta represents 0.0290 km/sec. The Betelgeuse and Bellatrix velocity values are just right to match their respective megalith markers. Alnilam is within 2 km/sec of its marker megalith, while Alnitak and Mintaka appear too slow. However, the radial velocity measures for Alnitak and Mintaka are considered very uncertain. Hipparcos doesn't even give a standard error estimate for the Alnitak radial velocity measurement. Better future observations may find that the Alnitak and Mintaka velocities do coincide with their megalith map values.

The Vega circle is the innermost, southernmost one. It is a little short of its marker megalith but within about a standard error of our knowledge of Vega's radial velocity. Also, Vega is the only one of these stars that is moving toward us rather than away from us. Because Vega's marker megalith is on autumnal equinox rather than vernal equinox, the same sign chart velocity may be seen as consistent with the other markers.

Yet a further level of significance of the megalithic map corroborates that the northern megalith placements do in fact indicate velocities.

Planetary systems map

In the Nabta megalith map so far, one megalith in each southern alignment represents the physical distance to that star, and one megalith in each corresponding northern alignment represents the radial velocity of that star. Also, each corresponding northern and southern pair of megalith alignments contain the same respective numbers of megaliths.

An exception to the rule of equal northern and southern numbers in each line appears to be the northern megaliths for the Orion's Belt stars that are shown as single megaliths whereas they should be double to match their respective southern megaliths. However, most of the actual megaliths are broken. Wendorf and colleagues did their best to analyze which megaliths could clearly be reassembled and which fragments were probably separate, resulting in the location measures used herein. Megalith A-2 (for Alnitak) Wendorf states was probably a double stone, and Megaliths A-1 and A-3 (for Alnilam and Mintaka) are also possibly double.

If we continue to look at the site as an astrophysical map, and note that each alignment has one primary stone representing the primary star plus secondary stones, the obvious question to consider is whether the other megaliths in each line may represent planets or secondary "companion" stars to the primary star.

This hypothesis cannot yet be tested against observations because modern astronomers are not yet able to observe the planetary and companion systems of these stars. Extrasolar planet detection techniques are proceeding rapidly apace though and some day soon we may be able to observe these systems.

However, we *can* test the physics of this hypothesis. If the hypothesis does apply to the stones, then the megaliths must be placed according to the astrophysical laws of planetary motion. The southern direct line of megaliths placements should represent the mean orbital distances of the planets or companion stars ("secondaries"), and their respective northern velocity marker megaliths should represent the mean orbital velocities of the secondaries. Thus, applying physics via Kepler's laws of planetary motion to the megaliths in each alignment can demonstrate whether or not the satellite megaliths might represent actual satellites of the primary stars.

Note that we are now going to do a kind of *reverse astrophysics*. When engaged in reverse engineering in business, a technologist is faced with a finished mechanical product but no documentation as to how the device was engineered. He tries to work backwards to discover the principles employed to produce it. We are faced with a prehistoric megalithic astrophysical map, and no documentation as to what it means (of course, that's the definition of "prehistory"), and we will work backwards to test what astrophysical principles might have been used to make the map.

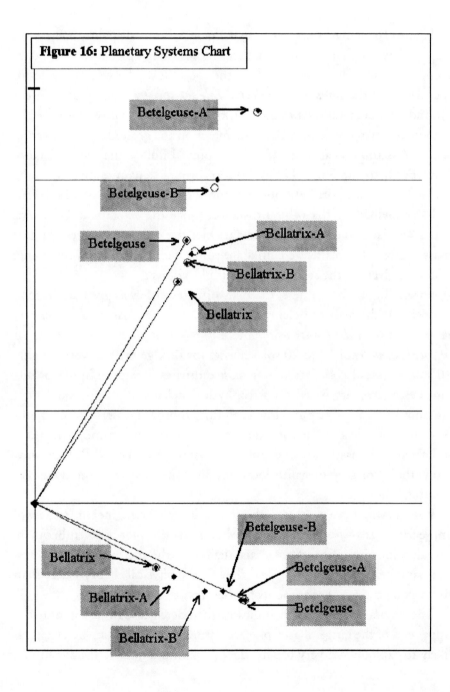

Figure 16: Planetary Systems Chart

Figure 16 shows very good agreement of the megalith map with Kepler's laws of planetary motion. This indicates that planets or secondary companion stars are represented by the megaliths. The velocity scale for the planets is the *same as the velocity scale for the stars* (one meter on the ground at Nabta represents 0.0290 km/sec). The distance scale for the planets/companions is such that one meter on the ground at Nabta represents 17 astronomical units. (An astronomical unit is the mean distance from Earth to the Sun.) The physics of orbital dynamics gives that the square of the mean orbital velocity is proportional to the mass of the central star divided by the orbital distance. Once the velocity and distance scales are fixed, there is only one variable: the star mass. Once the star mass is given, the planet velocities are computed directly from the planet distances given by the southern megalith locations.

For the figure shown here, the mass of Betelgeuse was given as 18 solar masses, and the mass of Bellatrix was given as 5 solar masses. *These masses, determined from the Nabta Map, are in very good agreement with the astrophysical estimates* of 12 to 20 solar masses for Betelgeuse, and very roughly 10 solar masses for Bellatrix. Star mass estimates, like pre-Hipparcos distance estimates, are based on astrophysical stellar evolution models that are only very roughly approximate. Future improved observational methods will probably allow astronomers to determine whether these megalith-indicated star masses are the precise physical star masses. But we now know the secondary megalith locations are consistent with the dynamics of orbital motion.

Readers who have worked with actual scientific or engineering data will appreciate that I was nearly floored when I saw this plot. Remember these megalith location markers are actual data from the site, constructed seven thousand years ago or longer, and they fit the physical theory better than many good modern experiments fit their theory the first time.

The Nabta megalith map indicates that Betelgeuse's mass is near the upper end of the range of astrophysical estimates. That means Betelgeuse is likely to explode in a very bright supernova, in the not too distant future, and then collapse into a neutron star. A smaller mass Betelgeuse would follow a different evolutionary track not leading to a compact neutron star.

Companion system plots for Alnitak, Alnilam and Mintaka are not shown because their companions are indicated comparatively distant from their primary stars. Their companions' relative velocities would be small and consistent with the megalith map showing the secondary velocity indicator megaliths close to their primary stars.

GALAXY MAP

The wealth of information contained in the megalithic astrophysical map compels further study. The rest of the Nabta Playa site, what little of it is excavated so far, analyzed in light of these findings, continues to yield more stunning finds. Given the understanding that the megalithic placements form a star and star systems map, this knowledge can be applied to understand the structures at the center of the megalithic complex.

The megalith alignments radiate out from one central point. At that central location, on the Nabta Playa sediments, a "Complex Structure" of several megaliths exists. One upright megalith marks the center of the radiant alignments and several other megaliths were placed to form an oval shape a few meters across. Several similar megalithic constructions, reportedly at least 30, exist in the vicinity of the central one. Archaeologists couldn't develop a clear understanding of what these megalithic structures are, and dubbed them simply "Complex Structures", and called the central one "Complex Structure A".

When they discovered these structures, archaeologists originally hoped they might be graves of elite Neolithic chieftains. They set out to excavate in hopes of finding bones and grave goods. Only two of the structures, Complex Structure A and Complex Structure B, have been excavated and a few others partially probed. No human or animal remains were found, nor any grave goods. Bizarrely, the surface Complex Structure megaliths mark only sculptures *carved onto the bedrock* 8 to 12 feet beneath the surface.

Complex Structure A, at the center of the whole system of star-aligned megaliths that radiate outward, contained an additional feature underground. Buried under the surface arrangement of large stones, and placed above the bedrock sculpture, was found another large (about 4,000 pounds) sculpture. This megalith sculpture is carefully and extensively sculpted into a complex geometric form. It is described by archaeologists as "vaguely resembling a cow". I will discuss the meaning of this "Cow Stone" in a future section, and now return to the bedrock sculpture.

In light of the meaning of the megalith alignments as an astrophysical, star and star systems map, the purpose of Complex Structure A becomes apparent. Because it is located at the center of the alignments to the star systems, logically it should in some sense represent our location from which we view the stars represented by the megalith alignments.

Following that logic, I first considered whether Complex Structure A might be a representation of Earth or the Sun. But those approaches leave the bedrock sculpture looking like nothing more than an oddly shaped rock.

Further pursuing the approach that it should represent our location reveals that it appears to be a sculpture of our Milky Way Galaxy, as it would be seen from the outside, viewed from the galactic north pole.

Our Sun's location is indicated in the proper place, to scale, and in the correct orientation on the galaxy sculpture by the location of the surface megalith that marked the center of all the radiating megalith alignments, continuing the surface megalith diagram system of marking vernal equinox heliacal rising, the direction from the Sun on the sculpture to the Galactic Center on the sculpture points to the actual direction of the vernal equinox heliacal rising of the Galactic Center as viewed from Nabta Playa. This is shown in Figure 17.

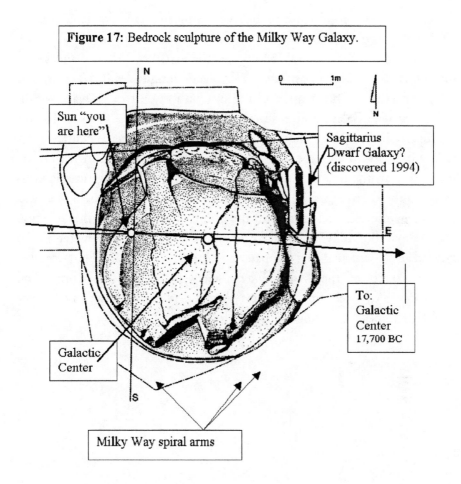

Figure 17: Bedrock sculpture of the Milky Way Galaxy.

The background of Figure 17 is fashioned after a drawing of the bedrock sculpture underneath Complex Structure A by M. Puszkarski, and the map with N-S and E-W coordinates mapped to the central point of the megalith alignments is by R. Schild and H. Krolik, from the *Holocene Settlement of the Egyptian Sahara* book. Over their map I've drawn the location of the Sun and the Galactic Center correlated to the direction of the vernal equinox heliacal rising of the Galactic Center in 17,700 BC, and to scale for the known distance to the center of the galaxy.

In the Milky Way Galaxy Sculpture, the east, north, west, and south directions correspond to astronomical galactic longitudes 0, 90, 180, and 270, respectively. The location of the sun with respect to its place within the Milky Way Galaxy and in scale and orientation to the Galactic Center is properly indicated by the surface standing megalith at the center of the radiating megalith alignments. The location of the star alignments radiating point, on the bedrock sculpture, off center on the sculpture, is indicated on the drawing of the sculpture as it was found in the field, and was marked by the archaeologists with the N-S and E-W lines in the drawing. The scale of the Milky Way Galaxy Sculpture is such that one meter on the ground at Nabta equals 5.4 kiloparsecs (a kiloparsec is 3,260 light-years).

The sketch of the Milky Way Galaxy in Figure 18, from "The Electronic Sky" web site, confirms that the Nabta Milky Way Galaxy Sculpture appears to be a good match to the actual data.

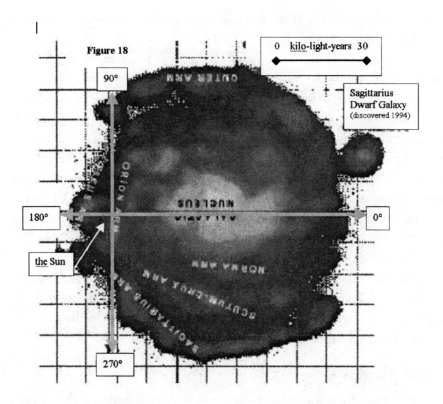

Figure 18

The Electronic Sky drawing is based on astrophysical data. Figure 18 is the Electronic Sky image rotated so that it is viewed from the galactic north pole with our sun to the left. The galactic longitudes are labeled, and our location "the Sun" is located at the center of the coordinate system as is the astronomical convention.

The Sagittarius Dwarf Galaxy, a recently discovered small satellite galaxy to our Milky Way may even be represented in the correct place on the sculpture as are the galactic spiral arms. In fact recent astrophysical analysis of the shape and location of the Sagittarius Dwarf Galaxy indicates that it more closely fits the Nabta sculpture representation than the

Electronic Sky drawing (Helmi and White, 2001, in *Monthly Notices of the Royal Astronomical Society*).

As seen from the central observing point, which represents the location of the sun on the Milky Way Sculpture, the sightline direction to the locus of the Galactic Center on the date of the Galactic Center's vernal equinox heliacal rising (circa 17,700 BC) is precisely as oriented on the sculpture. Further, the orientation of the actual galactic plane corresponded basically to the orientation of the sculpture on the ground on vernal equinox sunrise in 17,700 BC. (Motivated readers can verify this by considering the orientation of the north galactic pole in 17,700 BC [right ascension -91 degrees and declination 4 degrees]. This yields that at the time of vernal equinox heliacal rising of the Galactic Center, the galactic plane was roughly horizontal to the ground and in the same configuration as the sculpture.)

Astonishing as it may be, the bedrock sculpture underneath "Complex Structure A" at Nabta Playa appears to be an accurate depiction of our Milky Way Galaxy, as it was oriented astronomically at a specific time: vernal equinox heliacal rising of the Galactic Center in 17,700 BC.

Local galactic group map

Given that the purpose of the bedrock sculpture underneath Complex Structure A is a representation of the Milky Way Galaxy, the purpose of the other Complex Structures nearby becomes of interest. The Milky Way Galaxy Sculpture itself contains a clue to deciphering this next aspect of the megalith map. The sharp-cut vertical notches cut diagonally on either side of the sculpture define a vertical plane that correlates with a specific orientation of the Galaxy and the Galactic Center.

Complex Structure B:

The only other Complex Structure excavated so far, Complex Structure B, was chosen for excavation by the archaeologists because it is near the central Complex Structure A, about 45 meters away to the southeast. It was also chosen for digging because the surface megalith architecture of

Complex Structure B is extensive. Archaeologists were again disappointed to find no bones or grave goods at all. Instead, underneath eight feet of sediment they found another large oddly shaped lumpy sculpture cut directly onto the bedrock.

This sculpture is almost twice as big as the Milky Way Galaxy Sculpture and is an odd, lumpy, tilted, oval shape. It is probably a sculpture of the Andromeda Galaxy. The Andromeda Galaxy is a large nearby elliptical galaxy that is about twice as big as our Milky Way Galaxy, and together with our Milky Way it is the dominant member of the local group of smaller galaxies. The size of the Andromeda Galaxy Sculpture is to the same scale as the Milky Way Galaxy Sculpture.

The distance and direction to the Andromeda Galaxy Sculpture appear consistent with a projection from a coordinate system defined by the vertical notches cut into the Milky Way Galaxy Sculpture. That plane, based on the available drawings of the sculpture, appears consistent with the Milky Way Galaxy plane as it was on vernal equinox sunrise on the date of northern culmination of the Galactic Center, 10,909 BC. (The motivated reader can verify this by a study of the celestial coordinates of the Galactic Center and the galactic pole on that date. But we can only say "appears consistent with" in this case because only drawings of the Complex Structure B sculpture are available, and no detailed measures or models of it are available.)

The distance to the Andromeda Galaxy Sculpture appears consistent with a projection from a coordinate system rotated about 90 degrees within that plane, rather than a coordinate system defined by the plane and the actual direction to the Galactic Center at vernal equinox sunrise. The later coordinate system would seem to be the most natural representation, and the Nabta megaliths designers generally used intuitively elegant systems. So more measures and more excavations of the other "Complex Structures" will be needed to determine more certainly whether they represent the other nearby galaxies, and what the coordinate projection system is. The distance scale for the galactic map though is firmly fixed by the sculpture under Complex Structure A that unambiguously represents our Milky Way Galaxy.

SCALING LAW

So far, the Nabta Playa megalithic map has employed three distance scales: 17.0 AU/m for planets and companion stars; 0.799 LY/m for stars; and 17,600 LY/m for galaxies. The designers of the map needed to employ different scales to fit the different types of objects together on the map. Each type of object (planet, star, galaxy) is placed on the map according to the scale for that type. Different scales for different types of objects are called "scaling laws". Continuing with our astrophysical reverse engineering, to search for more information in the map at other scales, it is logical to look for a pattern in the scaling laws. If there is such a pattern, one might expect it to be a natural, nondimensional pattern. (Nondimensional meaning not depending on the use of an artificial measure like meters or feet.)

Nondimensionalizing the three scales (by expressing each in terms of meters on the ground to meters in the sky, thus eliminating the "meters") and then taking the natural logarithm, yields a sequence of whole numbers: (29, 37, 47). These are prime numbers. And the numbers appear to fit a pattern given by selecting the next largest prime number in a log-linear sequence. In such a sequence, the next scale in the map is given by 59.

COSMOLOGICAL COW STONE

This next scaling law gives a new scale of one meter on the ground at Nabta equals 2.87 billion light-years in the sky. This may yield another level of information in the megalithic map. In Complex Structure A, buried between the bedrock Milky Way Galaxy Sculpture and the surface, was found a megalith sculpted into a complex form. This large geometric sculpture has a "convex" outer surface, and was carefully placed with the outer surface (that appears close to spherical in the available drawings of it) facing away from the point in the Milky Way Galaxy Sculpture that indicates our location, the Sun. The sculpture also appears to have been placed at the proper distance such that the center of curvature of the sculpture is located at the Sun. Such a placement suggests a meaning for the sculpture other than a poor attempt at a cow.

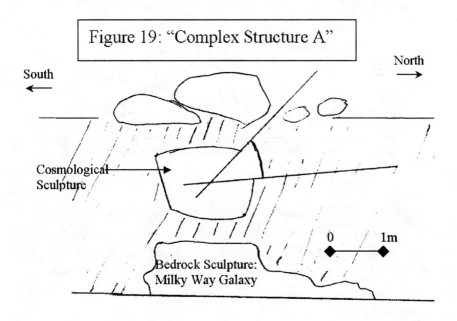

Figure 19: "Complex Structure A"

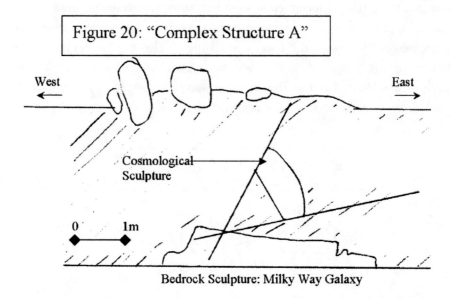

Figure 20: "Complex Structure A"

West ←

East →

Cosmological Sculpture

0   1m

Bedrock Sculpture: Milky Way Galaxy

Figures 19 and 20 are schematic drawings of cross-sections of Complex Structure A. Figure 19 is a south-north cross-section, and Figure 20 is a west-east cross-section. (These drawings are based on the archaeological drawing by M. Puszkarski, and the archaeological profile by R. Schild, F. Wendorf, S. Duncan and H. Krolik, from the book *Holocene Settlement of the Egyptian Sahara*.)

The "Cow Stone" sculpture, that I label as "Cosmological Sculpture", is shown as it was placed in the ground above the Milky Way Galaxy Sculpture. In Figure 20, I've added lines showing that the edges of the sculpture form a wedge that is centered on the location of the Sun in the Milky Way Galaxy sculpture. In Figure 19 I've added lines showing that the angular protrusion on the corner of the Cosmological Sculpture forms another wedge centered within the sculpture itself. These wedge line indicators that I've drawn are only approximate estimates based on the available drawings and data of the sculpture.

The distance from the Sun to the outer spherical edge of the sculpture, according to the scaling law, is about 6 billion light-years. This is highly suggestive that the sculpted stone is a representation of the cosmological "Big Bang" of creation. Possibly it represents the age of the universe at the time of the origin of our solar system (which is roughly 5 billion years old). Or possibly the designers of the map meant for the diameter of curvature of the sculpture, rather than its radius of curvature, to represent the current age of the universe—12 billion years. Or a third possibility is that it represents a current universe age of 6 billion years which would mean that our astrophysical models are in error by a factor of about 2.

In any case, the cosmological significance of the sculpture is further corroborated. The archaeologists describe the edges of this Cosmological Sculpture, which they speculate is a ritual depiction of a cow, as "polished smooth" and the spherical outer surface as "pecked smooth". Such a sculpted, mottled outer surface may plausibly represent the minute variations in the cosmic microwave background radiation as has been observed by the COBE satellite observatory. If interdisciplinary scientists are allowed to examine the sculpture closely, the plausibility of this aspect may be determined, and the significance of other markings on the stone could be studied.

More corroboration of the cosmological significance of this sculpture is found in its shape, size and orientation. Except for the angular protrusion on one corner, the four curved, angle cornered sides of the stone may match a shape defined by two lines of declination on the sides, and on the top and bottom by two lines of right ascension. This is shown in Figure 21 which is a drawing of such a declination window for the Galactic Center as viewed from Nabta Playa looking roughly east and up. Thus the shape of the sculpture, as it was placed facing outward from the sun location on the Milky Way Galaxy Sculpture, forms a "declination window" for the Galactic Center. The sculpture appears to have been carefully oriented such that this declination window subtends the full range of motion of the declinations of the Galactic Center. The sculpture marks the apparent motion of the Galactic Center throughout the full 25,900 year precession cycle. A line from the representation of the Sun to the actual Galactic Center passes though the window for a period of about 3 hrs every day, starting about 40 minutes after it rises above the horizon.

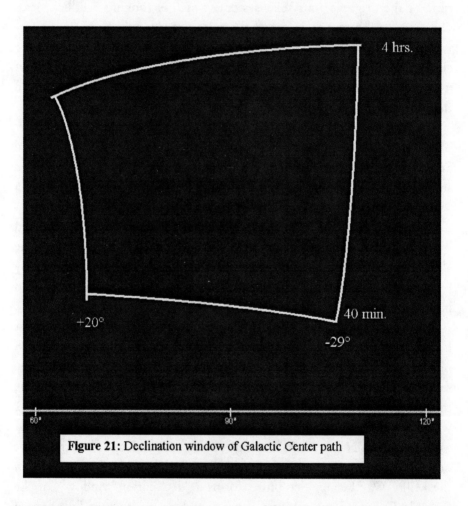

Figure 21: Declination window of Galactic Center path

The angular protrusion on one corner of the "Cow Stone" suggests further cosmological meaning. The two planes that border the protrusion define a wedge into the stone, and thus into the Galactic Center declination window. And the wedge converges at one point, defining a specific orientation to the Galactic Center on a specific date at a specific time. This implies a possible integration of meanings with specific Galactic Center orientations in the Giza plateau monuments discussed later in this book. Better measurements of the sculpture will determine the specific angles and dates and address the plausibility of this aspect.

A further corroboration of cosmological significance appears to be the orientation of the angular protrusion on the Cow Stone. It was placed such that, at vernal equinox heliacal rising of the galactic center in 17,700 BC, the protrusion pointed roughly in the direction of what astrophysicists call the "least velocity direction of the cosmic background radiation dipole". That means it points in the direction from which we came, in a cosmological sense, which can conceptually be interpreted as pointing toward the Big Bang of creation.

Planck scale physics map

Finally, if we apply the scaling law sequence once more, the inverse natural log of the prime number 79 would give a scale that could only represent objects much larger than the known universe. However, if we now reverse the scale from macro to microcosm, the scaling law sequence yields a scale that could represent extremely small things: 0.505 meters at Nabta equals one Planck Length.

The Planck Length is considered the fundamental length in physics. It is a fundamental constant of nature, and is the natural length that derives from three other fundamental constants of nature: Planck's constant, the Universal Gravitation constant, and the speed of light. At an almost unimaginably tiny 1.6×10^{-35} meters the Planck Length is the size smaller than which all known physics breaks down. At this scale, the Cow Sculpture (Cosmological Sculpture) is close to one Planck Length thick. So the Cosmological Sculpture may double as a Planck Scale Sculpture. This is suggestive that on the sculpture, or elsewhere in the map, may be contained information about Planck scale physics. For example, the angular protrusion on the Cosmological Sculpture is defined by two planes that are in a sense orthogonal to all of space-time. In the galactic portion of the map, when an orthogonal plane was introduced it represented an entirely new coordinate system. Similarly the angular shape and/or size of the orthogonal plane protrusion on the Cosmological Sculpture may represent information about physical dimensions orthogonal to space-time. Specifically it would be a natural way to represent aspects of multidimensional Planck scale physics such as how many physically meaningful hyperdimensions there are, eleven perhaps. The archaeologists describe the end of the sculpture opposite the angular protrusion as having "several cut marks and grooves". They claim those marks are "placed to control the flakes that shaped that end". In light of what this sculpture and the associated megalithic map really are, *the sculpture's surface should be carefully preserved* and interdisciplinary scientists and scholars should examine it for traces of more evidence.

PART III

Giza Monuments Galactic Zodiac Clock

Introduction to Giza study

The primary purpose of this book is to make available to scholars, scientists and interested readers the Origin Map discovery as given in *Part II*. The current part of this book presents the initial study that led me to investigate the Nabta Playa site in the first place. This study of the astronomy of the monuments on the Giza plateau is also another line of corroboration for the Nabta Origin Map. This Giza study should stand or fall on its own merits without affecting the integrity of the Origin Map find. That is why I will present the following study basically as it was initially presented, before I was aware of Nabta Playa. After this initial presentation, a discussion of unifying aspects of the Nabta Playa and Giza sites will be given.

This study of ancient skies over Giza, Egypt will show that the major monuments—the three Great Pyramids, their associated "valley temples", connecting causeways, and two sets of small "satellite pyramids"—have designed in their layout several congruencies with the stars of Orion's Belt around 11,000 BC. Also, the Giza monuments mark the location of the Galactic Center at its northern astronomical culmination circa 10,909 B.C. And the Khufu Pyramid's three star shafts simultaneously mark two ritually significant stars together with the Galactic Center around 2,350 BC, instead of with Orion's Belt as first indicated by Trimble and Badawy which aligned 150 years earlier. Further, the complex of monuments functions as a clock. This clock can be used to calibrate the start of the Zodiac to the northern astronomical culmination of the Galactic Center. And it marks a sequence of astronomical alignments that signify two Zodiac Ages each one twelfth of the approximately 25,900 year equinox precession period.

Many ancient megalithic monuments are known to have astronomical alignments. Deviations from precision of the true North alignments of pyramids in Egypt have been used to date the plan layout of the "Great Pyramid" of Khufu to 2,478 BC by assuming the pyramid designers used certain star sighting techniques for the alignment (Spence, in *Nature*). Bauval and others have noted that the three Great Pyramids of Giza appear to model the three stars of Orion's Belt, and the Sphinx facing due East suggests alignment to a vernal equinox sunrise. Because the vernal equinox sun rose in the constellation Leo most recently about 10,000 BC and the constellation Orion appeared in the southern skies of Egypt at vernal equinox at that time, Hancock and Bauval asserted the Giza complex layout represents that epoch, and its basic plan was set at that time, long before the three Great Pyramids were built on top of or replacing pre-existing monuments. These finds coincide with the geological studies by Schoch and West that indicate the Sphinx body was carved earlier than 5,000 BC, evidencing that some of the Giza plateau site significantly pre-dates the Great Pyramids built during Dynastic Egypt.

ORION'S BELT

To test Bauval's suggestion the three stars of Orion's Belt (Alnitak, Alnilam and Mintaka) are modeled by the layout of the three Great Pyramids of Giza, one needs a way to calculate the positions of the stars at that time. Professional quality programs like SkyMap Pro employ Julian Date, so they cannot calculate earlier than 4,712 BC, all they can be used for is to get a general idea of the positions of stars and how they move with precession. And the popularly available programs that do go further back in time don't accurately include effects like the motion of the ecliptic pole and the variation of Earth's obliquity which are important for accurate representations of such long past times.

So I had to create a program that could accurately calculate the positions of stars for tens or hundreds of thousands of years. A good method to do that was developed by French astrophysicist Berger, in the 1970's, so all one needs to do is put Berger's method into a usable computer code.

Calculating star locations

The apparent motions of stars are the result of the motions of Earth. Standard mathematical rotations (called Euler angle rotations) of the sky coordinates give the ancient locations of stars. Star proper motions, the physical movements of the stars with respect to each other (stars are all zipping around higgledy piggledy), are usually small compared to precession motions, but are important for a few cases, and so I include star proper motions in most of these calculations. Three basic rotations are needed, one for Earth obliquity which can vary by a few degrees, another for motion of the ecliptic pole which can also vary by a few degrees, and one for the general precession which describes a circle inclined about 23 degrees moving at variable rate. Even the general precession moves at a variable rate such that the precession cycle currently moves at a rate of 25,900 years but past and future cycles change from that by several hundred years. These three things, precession rate, ecliptic pole orientation, and obliquity, all are determined by the physical forces on Earth and so can be calculated from first principles of physics. All three rates change through time and all oscillate with variable periods about very long term average values. But for short periods of time of about a million years these motions can be written down very accurately, mathematically in closed form as Berger did. All one needs to do is translate Berger's mathematical equations into a sequence of instructions that a computer can do accurately. The actual calculation method is described in Appendix A.

Interestingly, over extremely long periods of time like hundreds of millions or billions of years even these steady motions exhibit some signatures of chaotic motion as has been shown by physicist Jack Wisdom (e.g. Touma and Wisdom 1994; Quinn et al. 1991) via computer simulations. And of course over those extremely long periods of time one also has to worry about niggling things like "perturbations" due to giant asteroids slamming into Earth, but that's a story for another book and not of concern to this one.

When did the pyramids match?

Bauval claimed that the Orion's Belt stars' angle in the sky matched the Pyramids' orientation on earth at a vernal equinox sunrise when Orion's Belt was at southern culmination in 10,500 BC. There are two ways to test that idea: 1) move the precession calculation back until the sky angle matched the ground angle *when the stars were on the meridian*; or 2) move the precession calculation back until the sky angle matched the ground angle *when the sun was at vernal equinox sunrise*. I did both of those tests.

First, I precessed the sky to when the angle of Orion's Belt in the sky, measured in altitude-azimuth coordinates, matched the layout of the pyramids on earth, measured in ground cardinal coordinates, exactly at the time of meridian transit (daily southern culmination) of the Belt stars. This turns out to be 11,772 BC, not 10,500 BC. This is when the ground plan map of the Great Pyramids, slid straight up into the sky in the southerly direction, matched the star positions in the sky.

Figure 22 shows a plan view of the Giza Plateau monuments. These plan view outlines are from accurate digital bitmaps of the Giza plateau monuments from Lehner (1997), and also available at the University of Chicago Oriental Institute Giza Mapping Project website. Small squares mark the centers of the three Great Pyramids. (See photographs of the Great Pyramids and Sphinx in Appendix C of this book.)

N

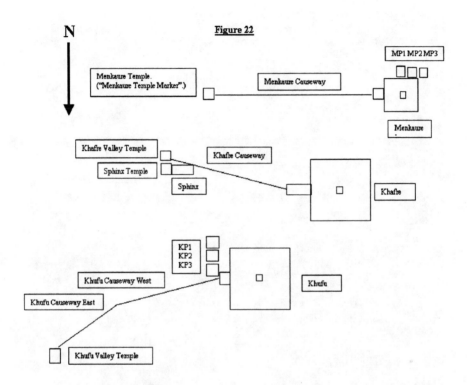

Figure 22

Figure 23 shows 11,772 BC vernal equinox Alnilam transit of the meridian. The plan view schematic of the monuments is plotted over a sky view image of Orion's Belt stars when Alnilam was on the meridian when their angle in the sky matched the ground angle of the pyramids. Neither view is rotated or stretched, only matched by size. The date was found by matching the angle of the line connecting Alnitak and Mintaka to the angle of the monuments, using the long-term accurate Earth celestial pole motion calculations. As can be seen, the pattern matching of the stars to the pyramid centers is quite good.

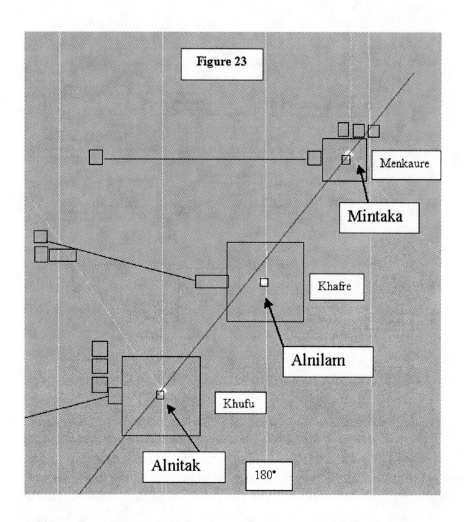

Figure 23

Menkaure

Mintaka

Khafre

Alnilam

Khufu

Alnitak

180°

Second, I precessed the sky to a vernal equinox date such that at the moment of sunrise the Orion's Belt in the sky, measured in Altitude-Azimuth, matches the layout of the Pyramids on earth, measured in ground cardinal coordinates. That turns out to be 9,420 BC.

Figure 24 shows the 9,420 BC vernal equinox sunrise match. The plan view schematic of the monuments is plotted over a sky view image of

Orion's Belt stars when their angle in the sky matched the ground angle of the pyramids. Neither view is rotated or stretched, only matched by size. The date was found by matching the angle of the line connecting Alnitak and Mintaka to the angle of the monuments, using the long-term accurate Earth celestial pole motion calculations.

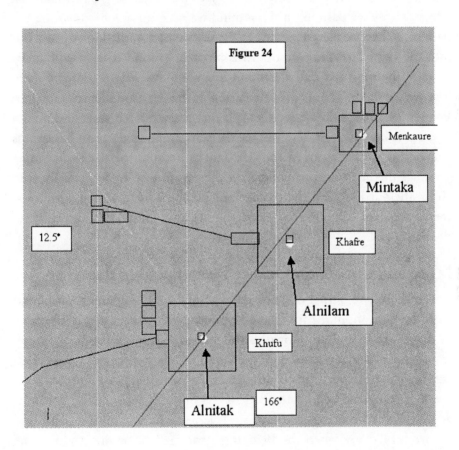

Figure 24

Orientation of the correspondence

Much has been made in debunking circles about a flippant comment by a planetarium director to the effect that the correspondence between the three Orion's Belt stars and the three Great Pyramids is "upside down" and so invalid. That is a silly claim. We live between ground and sky. When we view a pattern on the ground we necessarily look down on the pattern. When we view a pattern in the sky we necessarily look up and out into the sky to see the pattern. To compare a ground pattern with a sky pattern, the most natural, direct and unambiguous way to compare them is as we see them. That is the orientation of the correspondences shown in Figures 23 and 24, as we see them. If one goes to a toy store and buys a celestial sphere though, one will see the stars are indeed drawn backwards and not as we see them in the sky. This is because the toy maker must draw all the stars on one sphere, as they would be hypothetically seen "from the outside". But there is no "outside the stars" and that is not how we see the stars. Ground to sky pattern matching should be done as we see the patterns.

More alignments on Orion's Belt matching dates

On those two vernal equinox dates several other alignments occurred with the Giza monuments. These alignment events are listed in the table in Appendix B, and are shown in the following figures. Figure 25 shows the alignments that occurred on the 9,420 BC date when the Orion's Belt star matched the pyramids configuration at vernal equinox sunrise. First, the Belt stars rose behind the rear of the Sphinx. Second, the Belt stars angle matched the pyramid layout at vernal equinox sunrise. And that occurred precisely when the Belt stars happened to be directly over the Menkaure Valley Temple. This is significant because the congruence date and location was determined only from the pyramid locations and from the star movements, independent of the Menkaure Valley Temple. Third, the Orion's Belt stars set just inside of the three smaller Menkaure Satellite Pyramids that signify the Belt stars.

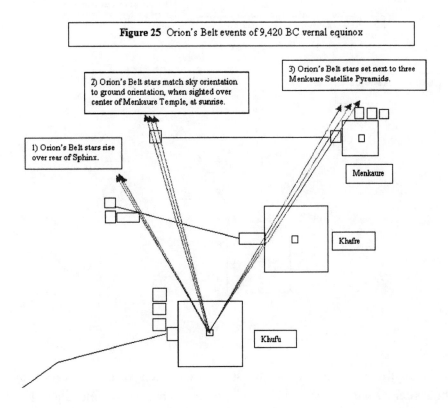

Figure 25 Orion's Belt events of 9,420 BC vernal equinox

2) Orion's Belt stars match sky orientation to ground orientation, when sighted over center of Menkaure Temple, at sunrise.

3) Orion's Belt stars set next to three Menkaure Satellite Pyramids.

1) Orion's Belt stars rise over rear of Sphinx.

Menkaure

Khafre

Khufu

Figure 26 shows the alignments that occurred on the 11,772 BC date when the Orion's Belt stars matched the pyramids configuration at meridian transit. First, the Belt stars rose over the rear of the Sphinx. Second, the Belt stars angle matched the pyramid layout at meridian passage. Third, the Orion's Belt stars set just inside of, beginning to set over, the three smaller Menkaure Satellite Pyramids that signify the Belt stars.

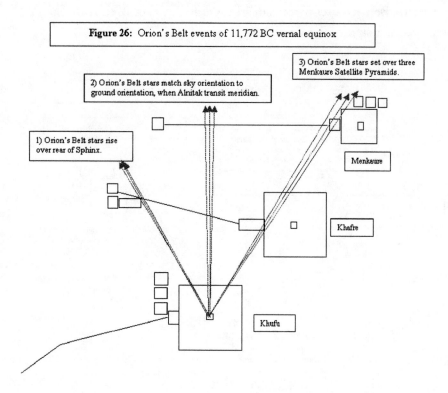

Figure 26: Orion's Belt events of 11,772 BC vernal equinox

All alignments described in this study are as seen from the location of the center of what is now the Khufu Pyramid, except for one alignment from the center of the Menkaure Pyramid as will be noted. The ground sight line angles are calculated from the accurate monument measures listed in the table in the Appendix B.

These alignments are calculated from ground level perspective. So this study envisions a pre-existing layout with small monuments or marker megalithic structures at the same locations, but not of the gigantic sizes, as the current Pyramids. (Ancient megalithic sites around the world often are found to be built on top of previous smaller monuments, sometimes thousands of years older.) Even so, the alignments calculated here would also substantially apply to sightlines from the tops of the current monuments as they have been since 2,400 BC.

The centers of the monuments are used for these sight angles and alignments, except where noted. The calculated times of these events yield star location dates accurate to within a few years. The alignment dates in this study are also shown in the table in Appendix B.

Regarding the alignments in Figures 25 and 26, there is no unambiguous way to calculate a probability that the Giza monuments were placed randomly, and these congruencies occur by happenstance. But it appears obvious that the probability is low that these alignments, combined with the close representational matching to the Orion's Belt stars and to the Sphinx-Leo matching, could be purely the result of random or unintended monument locations.

Regarding daylight vs. night events, these are stellar alignments, rather than planetary, and thus the correspondences and alignments appear essentially the same on, and can be viewed at, autumnal equinox shifted by 12 hours from the vernal equinox times so that daylight events become night events. For instance the setting of the 3 Orion's Belt stars over their three marker pyramids, in 9,420 BC, would have been visible on autumnal equinox in that year.

[Note: It has been claimed that Fourth Dynasty Egyptians "didn't know about" the constellation Leo. Even if that were the case, it is not relevant here. These alignments indicate planning by peoples vastly predating Dynastic Egypt. That early culture likely informed the Zodiacal lore of later cultures such as the Babylonian-Greek from whom came the modern Western Zodiac figures. Furthermore, the near-linear figure of the three Orion's Belt stars is an obvious sky feature, or "asterism", and does not depend on a constellation representation to be noticed by any sky watcher. This indicates designed alignments even if the Sphinx were not a lion and even if Orion were not a constellation at all. The Dynastic Egyptians who constructed the Great Pyramids circa 2,500 BC need not have known the full meaning of the pre-existing monuments, except that they were sacred and very precisely aligned to some kind of astronomical events.]

GALACTIC CENTER

Given the representation of the Orion's Belt stars in 9,420 BC and 11,772 BC, the question arises: why monumentalize that epoch? If the purpose was simply to mark the southern culmination of Orion's Belt, which happened around 10,600 BC, the correlation dates are off—the design seems too complicated to only be marking Orion's Belt southern culmination.

Some ancient historical traditions speak of the time around 11,000BC as an important time in human history. I discuss this point later. It is fruitful to consider the epoch from a purely astronomical standpoint first.

The Orion's Belt stars are near their minimal (southern) culmination in the sky around the matching dates. Culmination is when stars reach their highest or lowest arcs in the sky. Southern culmination is when a star transits the meridian at the time of the vernal equinox sunrise, and is defined as when celestial longitude or "Right Ascension" equals 18 hours (6 hours for northern culmination).

The two ground-sky orientation matching dates are *not* equal to the culmination time of the Belt stars as Bauval suggested—the two dates *bracket* the culmination time, and the dates span just more than 30 degrees of general precession (one twelfth of the Zodiac, which we now recognize as a Zodiac Age). The dates also bracket the Galactic Center northern culmination. Galactic Center northern culmination is when it reaches its highest point in the sky, and when the north spin axis of Earth comes closest to a line connecting the Earth and the Galactic Center. The Galactic Center culmination occurred between the two Orion's Belt star correspondence dates, in 10,909 BC. A study of the apparent motion of the Galactic Center, at that epoch, reveals the following alignments of the Giza monuments with the Galactic Center. I argue that Galactic Center culmination is signified by the complex of monuments, and should be the calibration point for the Zodiac calendar.

Zodiac clock markers

To investigate whether the monuments mark one Zodiac Age around the Galactic Center calibration point, look at the sky objects' motions 30 degrees of general precession before and after the Galactic Center culmination. The table in Appendix B shows declinations and right ascensions for the Orion's Belt stars, the Galactic Center, and the rear and paw stars of the constellation Leo on the dates of interest. Figure 27 shows the Orion's Belt stars' rising and setting azimuths and the Galactic Center rising azimuths, on the Giza plateau, on the dates 30 degrees of general precession before and after the Galactic Center culmination date.

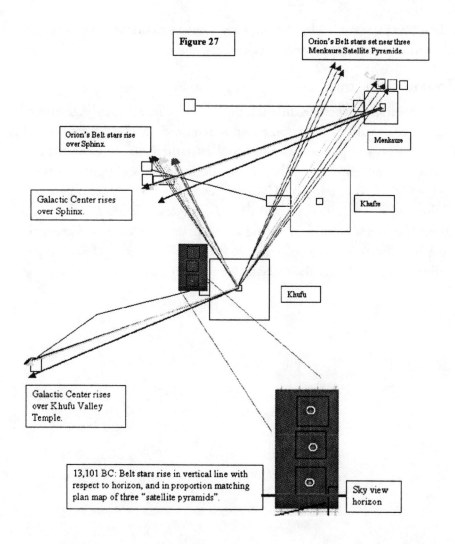

Figure 27

Orion's Belt stars set near three Menkaure Satellite Pyramids.

Orion's Belt stars rise over Sphinx.

Menkaure

Galactic Center rises over Sphinx.

Khafre

Khufu

Galactic Center rises over Khufu Valley Temple.

13,101 BC: Belt stars rise in vertical line with respect to horizon, and in proportion matching plan map of three "satellite pyramids".

Sky view horizon

In Figure 27 each set of arrows represents three dates: 30 degrees general precession before Galactic Center culmination (13,101 BC); Galactic Center culmination (10,909 BC); and 30 degrees general precession after Galactic Center culmination (8,707 BC).

On 13,101 BC, the Belt stars begin rising over the center of the Sphinx and setting over the center of the three Menkaure satellite pyramids, with the rising and setting azimuths passing inside the markers, indicating the approach of the calibration time. Also, the Galactic Center begins rising over the Khufu Valley Temple, and over the center of the Sphinx as seen from Menkaure. As years pass, these rising and setting azimuths pass inside their marker monuments to maximal positions indicated by the arrows, at Galactic Center culmination. The rising and setting directions then pass back out of the marker monuments 30 degrees of general precession (one Zodiac Age) later as shown by the arrows.

The inset in Figure 27 illustrates that around the 13,101 BC date the Belt stars rise above the horizon in a vertical line and in proportion to the plan map (north-south axis to vertical) of the three "satellite pyramids". Throughout the following dates the Belt stars rise angle with respect to the horizon increases, passing through matches with the upper causeway angle, the rising sight line to the Galactic Center, and then the lower causeway angle.

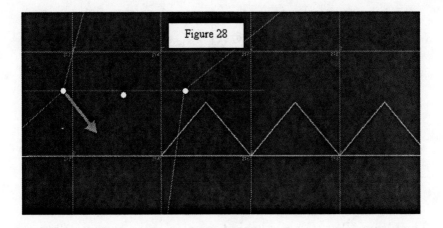

Figure 28

Figure 28 shows the Orion's Belt stars as they were seen in 13,101 BC, and again around 8,700 BC, setting in a horizontal line over the three

Menkaure satellite pyramids that represent them. The arrow shows the direction of motion.

Galactic Anticenter marked in sky map

Figure 29 shows the alignment events of 11,772 BC vernal equinox.

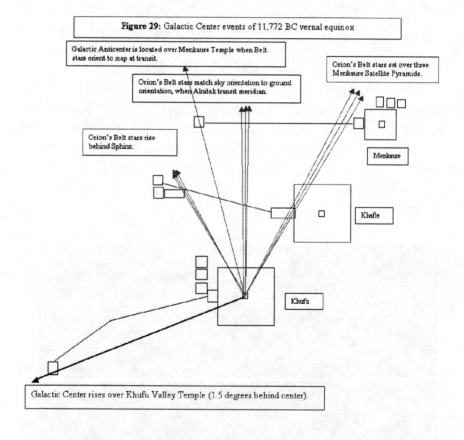

Figure 29: Galactic Center events of 11,772 BC vernal equinox

Galactic Anticenter is located over Menkaure Temple when Belt stars orient to map at transit.

Orion's Belt stars set over three Menkaure Satellite Pyramids.

Orion's Belt stars match sky orientation to ground orientation, when Alnitak transit meridian.

Orion's Belt stars rise behind Sphinx.

Menkaure

Khafre

Khufu

Galactic Center rises over Khufu Valley Temple (1.5 degrees behind center).

The Orion's Belt stars rise behind the rear of the Sphinx. When they transit the meridian they correspond to the Giza plan map on the ground.

And at that time the Galactic Anticenter is located over the Menkaure Valley Temple. This establishes another ground-sky map correspondence.

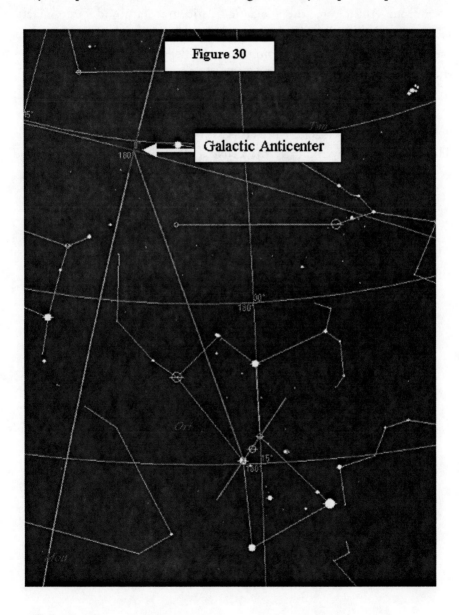

Figure 30

Figure 30 shows the same direction sight line as in the ground view on the 11,772 BC matching time, projected into a sky view. The sight line goes from Alnitak to the Galactic Anticenter (antipode to Galactic Center), at time of first sky-ground correspondence when Alnitak transits meridian. The Galactic Anticenter is at 165 degrees azimuth, directly over the Menkaure Valley Temple viewed from the center of Khufu, the primary observing point in this study. This sky view further corroborates that a sight line from Khufu, through the Menkaure Valley Temple marker, goes direct to the Galactic Anticenter in both the ground and sky views. This simultaneously indicates the Galactic Anticenter in the ground and sky maps.

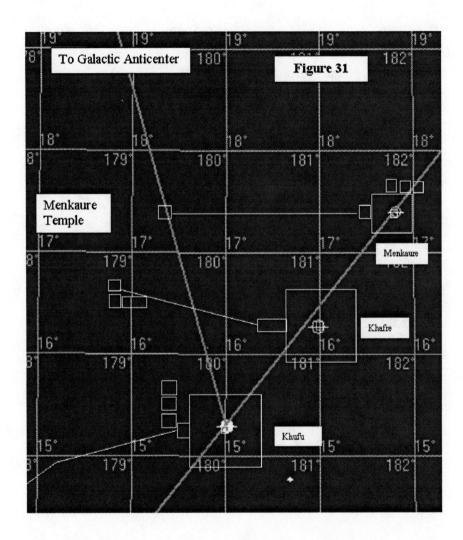

Figure 31 is a blowup of the central region of Figure 30 with an overlay of the schematic ground map of the monuments. It shows the first Belt stars correspondence to the Great Pyramids at vernal equinox, and the Menkaure Valley Temple marker sight line direct to the Galactic Anticenter.

Figure 32 shows that on the 9,420 BC alignment date the Galactic Anticenter transited the meridian with the Belt stars (actually with Mintaka, see the table in Appendix B).

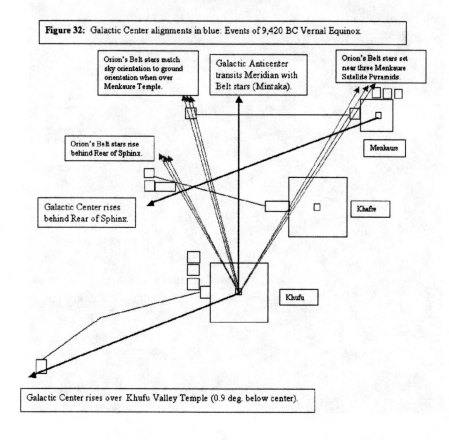

Figure 32: Galactic Center alignments in blue: Events of 9,420 BC Vernal Equinox.

Orion's Belt stars match sky orientation to ground orientation when over Menkaure Temple.

Galactic Anticenter transits Meridian with Belt stars (Mintaka).

Orion's Belt stars set near three Menkaure Satellite Pyramids.

Menkaure

Orion's Belt stars rise behind Rear of Sphinx.

Galactic Center rises behind Rear of Sphinx.

Khafre

Khufu

Galactic Center rises over Khufu Valley Temple (0.9 deg. below center).

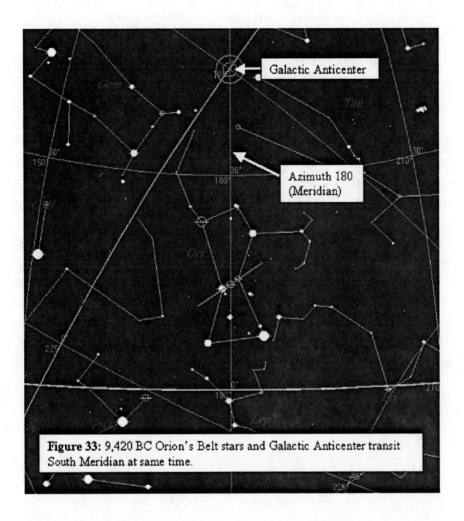

Figure 33: 9,420 BC Orion's Belt stars and Galactic Anticenter transit South Meridian at same time.

Figure 33 illustrates the further marking of the Galactic Anticenter by the monuments by its simultaneous transit of the meridian with the Belt stars on the vernal equinox sunrise matching date, 9,420 BC.

Star shafts align to Galactic Center

The shaft alignments in the Khufu pyramid further confirm the Galactic Center signification of the monuments. There are four meridian aligned shafts in the Great Pyramid of Khufu. No other major pyramid has meridian shafts. Recently, three of the shafts have been measured with great precision (Gantenbrink, 1999). These have been associated with alignment to culmination of the stars Alpha Draconis (Thuban), Sirius, and Alnitak (Zeta Orionis in Orion's Belt) at the time of construction of the Khufu pyramid. Those orientations would coincide with Dynastic Egypt mythology because Sirius and Orion were ritually significant, associated with the primary deities Isis and Osiris respectively, and Thuban was the pole star at that time and was ritually significant.

Alnitak did align with the King's chamber southern shaft near the date of cardinal alignment determined by Spence. But both Sirius and Thuban aligned more than 150 years *after* Alnitak aligned, as is shown in Figure 34. This suggests that if the Spence date was indeed the cardinal alignment date (plan layout date) for the Khufu Pyramid, then the 150 years later date of simultaneous Sirius and Thuban alignments could have been an intended ritual *usage date*, for use after the pyramid was constructed.

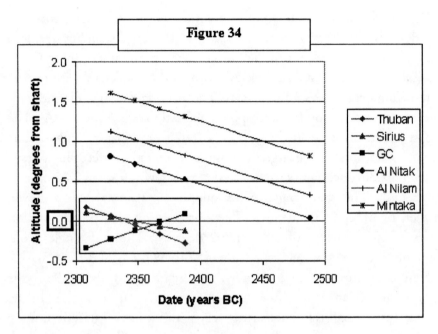

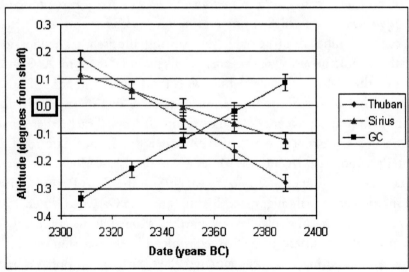

Figure 34 illustrates the Khufu pyramid "star shafts" alignments as a function of time. Thuban is with respect to the King's chamber northern shaft. Sirius is with respect to the Queen's chamber southern shaft. GC is the Galactic Center. GC, Alnitak, Alnilam, and Mintaka are with respect to the King's chamber southern shaft. The altitude data include the standard average correction for atmospheric refraction (see Appendix A). The largest refraction is small though, only 0.025 degrees for Thuban. The error bars are equal to the largest atmospheric refraction effect plus an estimated possible uncertainty to the celestial pole giving a small 0.05 degrees uncertainty. The shaft angle errors quoted by Gantenbrink are even smaller, of order 0.001 degrees. The star position and proper motion errors are also smaller (of order 0.002 degrees). The shaft angles from Gantenbrink, in degrees elevation, are (45.00, 39.6078, and 32.023).

Stars cycle through their culmination ranges during the 25,900 year equinox precession cycle. Thus the fact that some star culminates in the sight of a given shaft is not necessarily very significant. Two stars of primary ritual significance *passing culmination alignment with fixed meridian shafts at the same time* is very likely nonrandom. *Three objects* of primary ritual significance aligning with (the only three existing) precisely aligned meridian shafts, simultaneously, are extremely unlikely to be random. As Figure 34 shows, the Galactic Center aligned with the King's chamber southern shaft and Thuban and Sirius aligned with the other two shafts simultaneously to within measurement error and within 20 years of each other. That simultaneous alignment occurs 150 years after the Spence date for the Khufu Pyramid base layout, the time when the last Orion's Belt star, Alnitak aligned. This indicates a *functional date* for the Great Pyramid with triple simultaneous shaft alignment, highlighting the Galactic Center, 150 years after its construction plan layout. Further, a fourth (Queen's chamber northern) shaft was measured by Gantenbrink with less precision certainty because his measuring robot encountered physical barriers. But the best value from Gantenbrink (39deg 07min 28sec) aligns it with the ritually significant star Kochab in 2,368 BC—essentially simultaneously with the Galactic Center and the other two stars.

Zodiac clock

The two primary Orion's Belt star alignment dates (11,772 BC and 9,420 BC) are separated by 2,352 years and 32 degrees of general precession. Those two dates bracket the Galactic Center culmination alignment date as well as the Orion's Belt culmination dates. This suggests that the Giza monuments, as well as marking the date and location in the sky of the Galactic Center maximum culmination, function as a clock marking the passage of the Zodiac, the duration of a Zodiac period (30 degrees of general precession), and a calibration to the zero point of the Zodiac: culmination of the Galactic Center.

Other sequential alignments mark the Zodiac clock calibration points. Throughout the Galactic Center culmination epoch the Orion's Belt Stars rise over or near the Sphinx as seen from Khufu, and the Galactic Center rises over or near the Sphinx seen from Menkaure. And the Zodiac age of Leo is signified by the risings progressing from the rear of the Sphinx forward to its paws, as the Sun in vernal equinox passes from the constellation lion's rear forward past its paws.

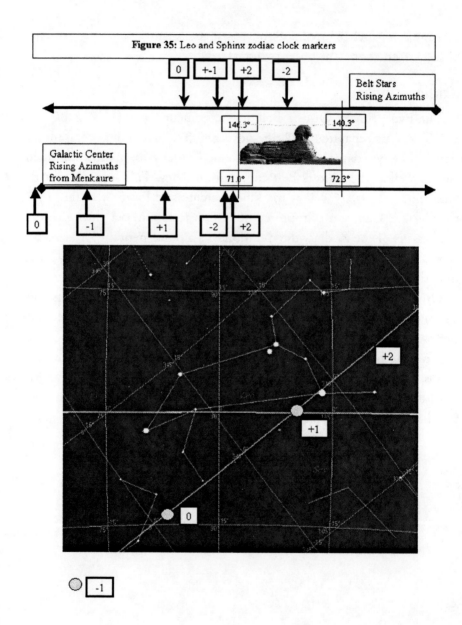

Figure 35: Leo and Sphinx zodiac clock markers

Figure 35 illustrates the Leo and Sphinx Zodiac Clocks. The figure shows the Orion's Belt rising sight lines with respect to the Sphinx; Galactic Center rising sight lines with respect to the Sphinx, and sun locations with respect to Leo, on the four vernal equinox clock dates. Clock Marker 0 calibrates the beginning of Zodiac Age Leo, with Galactic Center culmination, and with the start of the whole precession cycle. Clock Markers (-2, -1, +1, +2) respectively correspond to (13101, 11772, 9420, 8707) BC as in the table in Appendix B. The directions indicated in Figure 35 are nominal horizon rising directions. After rising, both the Galactic Center and the Belt stars move forward over the Sphinx, so at an elevation about one Sphinx length, all markers would move forward of order one Sphinx length, further corroborating the directions.

In the year of maximum culmination of the Galactic Center, the Orion's Belt stars rise behind the rear of the Sphinx, the Galactic Center rises behind the Sphinx seen from Menkaure, and the Sun begins the Zodiac Age Leo just behind the rear of the Lion constellation in the sky. The vernal equinox sun locations on these dates are plotted in Figure 35, using the right ascensions of Denebola and Omicron Leonis—the Lion constellation's rear and paw stars.

The Giza monuments thus mark the culmination of the Galactic Center as the calibration for the Zodiac, in several ways:

- Simultaneous movement across the Sphinx of the rise of Orion's Belt and the Galactic Center;

- Movement of Orion's Belt stars congruence from meridian to over Menkaure Valley Temple, marking about 30 degrees of general precession;

- Motion of setting of Orion's Belt Stars over three representational Menkaure satellite pyramids, in and out of the pattern over two Zodiac Ages;

- Motion of rising of Galactic Center through Khufu Valley Temple, in and out over the two Zodiac Ages;

- The Galactic Anticenter is marked in the corresponding sky map and ground sight lines;
- Simultaneous alignment of the Galactic Center, Sirius, Thuban and Kochab with the four ritual shafts of the Great Pyramid.

GIZA RESULTS

These data strongly indicate the following.

- The monuments function as ground and sky maps to signify the time and location in the sky of the culmination of the Galactic Center.
- The monuments function as a clock to mark the passage of the Zodiac Age of Leo, and to calibrate the start of the precession cycle to the culmination in the sky of the Galactic Center.

When were the monuments designed? The data evidences that the Giza plateau monuments design plan was either created more than 12,900 years ago and developed over a several thousand year period; or it was created by a people with astronomical calculation and conceptual design abilities rivaling our own, at some time before or contemporary with the large construction events at Giza around 2,400 BC.

GIZA DISCUSSION

Prevailing theory of Egyptologists seems to be that cultures previous to Dynastic Egypt could not have organized the planning and building of monuments and therefore evidence for such previous cultures should not be considered. But in a rigorous science such a methodology would be upside down. Data and observation should have precedence over theory. The data presented in this part of this book are essentially *observational* archaeoastronomy. Generally in a healthy science theory should endeavor to accommodate, not automatically to dismiss, new observations.

In this case, the observation of monuments that mark the time around 10,909 BC in relation to astronomical cycles is not out of context with ancient cultures' documented cosmologies. Some ancient cosmological references do state that the epoch marked by the Giza monuments was a primary turning point in human cultural development, and that it coincided with a cyclic astronomical aspect.

Vedic cosmology

There are several examples of ancient cultures that allude to prehistoric human high cultures much more ancient than the generally assumed date of the beginning of history (meaning the beginning of writing) around 3,500 BC. As one example, Vedic scholar S. Yukteswar published an analysis of the Vedic yuga system in 1893. Yukteswar noted that the original Vedic description of the yuga ages gave the yuga durations in 2 sets of four periods with ratios 4:3:2:1 (4800:3600:2400:1200) adding up to 24,000 years. Modern translations of the Vedas insert the word "divine" before "years" to create 360 times 24,000 (or 8,640,000) years as the full yuga epoch. The modern changing of "years" to "divine years" thus allows the denigration of the ancient Vedas to merely symbolic fantasy that can't have anything to do with actual human cultural history.

Yukteswar however suggests that the original Sanskrit Veda was accurately referring to years not "divine years" and was meant to coincide with precession of the equinox. Further, Yukteswar placed the zero point (the highest, most spiritually evolved time between the two 4,800-year Krita Yugas) at 11,501 BC. According to Yukteswar, very ancient Vedic astronomy conceived of planets revolving around the sun, and "*the sun also has another motion by which it revolves around a grand center called Vishnunabhi, which is the seat of the creative power, Brahma, the universal magnetism. Brahma regulates dharma, the mental virtue of the internal world...From 11,501 BC,...the sun began to move away from the point* of its orbit nearest to the grand center toward the point farthest from it, and accordingly the intellectual power of man began to diminish." Because the Great Pyramids and Sphinx mark dates around 11,500 BC, and especially because the monuments also highlight the northern culmination (or "nearest point" to the north pole) of the Galactic Center, certainly a "grand center around which the sun revolves", Yukteswar's insights and the original Vedas deserve more consideration.

Yukteswar claimed a very ancient and accurate yuga calendrical counting method, that was rigorously based on astronomy, was essentially forgotten by around 3,000 BC. When Indian Rajas attempted to revive the counting method around 600 AD, according to Yukteswar, the appointed astrologers recognized that the literal Vedic translation would mean the world was descending into a Kali Yuga (the least mentally evolved state of man) and the astrologers were uncertain how to calibrate the counting system because of their ignorance of astronomy. So to make their pronouncements more palatable the court astrologers asserted that the Vedas meant "divine years" not "years" such that our Kali Yuga would be 432,000 years long. In that case the yuga condition of modern man essentially wouldn't change in historical times. Those early modern scholars, working in an age that Europeans still call the beginning of the "Dark Ages", created that interpretation, according to Yukteswar, to avoid the depressing conclusions of a direct translation of the Vedas.

Yukteswar's calibration of the system was a bit inaccurate because he was unaware of the accurate astronomical equinox precession period and he assumed it to be exactly 24,000 years. But his placing of the most highly evolved consciousness point near to the 10,909 BC Galactic Center calibration we now find at Giza is interesting, given that Yukteswar himself seems to have had no concept of the astrophysical Galactic Center or knowledge of its location or its astronomical culmination.

Stencel et al. (1976) showed that at least one other major ancient monument complex encodes both the Vedic yuga system and precise astronomical alignments: Angkor Watt. Stencel et al. assumed the usual 360 "divine year" multiple in their yugas, but all their findings apply equally well if scaled down by 360 and could be interestingly revisited in the new light of the new evidence presented in this book.

Note that while traditional Vedic yuga interpretations name Kali Yuga as the age of the least sensitive and most materially-oriented human mind, a sort of "Age of Dullards", it need not necessarily be considered a "bad" time. It could also be seen as a time of development of mind excellence in the material arts and sciences, as opposed to Krita Yuga's development of mind excellence in the spiritual arts and sciences. And in fact such an interpretation may coincide with the brilliant and wonderful technological and material scientific advances of modern times.

The above two examples, yugas according to Yukteswar, and Angkor Watt according to Stencel et al., are from Vedic cosmology. But similar cosmologies of ancient cycles exist in other systems including those associated with the "western religions" as in the Bible of course.

Corroboration in Ancient Egyptian art?

Figure 36: Thebes Valley of the Kings Tomb Paintings.

36a

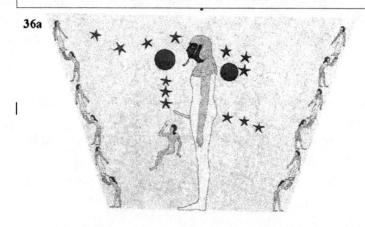

36b

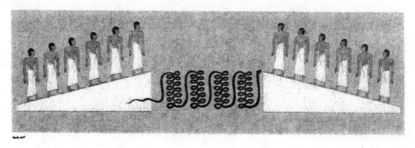

Some of the meanings of the Giza monument alignments may be expressed in known Ancient Egyptian art. Figure 36 shows two paintings, reproduced by the Napoleonic expedition of 1799, from a tomb in the Thebes Valley of the Kings. Fig 36a shows a pharaoic figure flanked by four sets of stars. Lower front of the pharaoh are three stars in the configuration of Orion's Belt rising vertically as they did around the 11,000 BC epoch; lower back of the pharaoh is Orion's Belt setting nearly horizontally as Orion's Belt did at that time. Upper front of the pharaoh are four stars with an important non-stellar sky object, upper back of the pharaoh are three stars with another important non-stellar sky object. Orthodox scholars tend to associate these disks with two representations of "the Sun disk". But these two may also represent the Galactic Center and Anticenter sky regions as they appeared at the same time as the Orion's Belt configurations. A queen figure looks and points up at the Galactic Center. The pharaoh's excitation could possibly signify a generative or creative time. Flanking that central scene are six ascending and six descending human figures representing the twelve Zodiac Ages, six ages of ascending mental sensitivity and six descending. *By the time this painting was created, for the 20th Dynasty pharaoh Sethnakht circa 1186 BC, an essentially modern culture was present. So I am not claiming that the painting iconography directly dates to the time of Galactic Center culmination; the painting may express small bits of cultural knowledge that partially survived to the time of creation of the painting.*

Figure 36b shows six human figures on an ascending slab and six on a descending slab, facing a central snake coiled neatly in four repeating sets of twelve coils each. Snake often represents both creative energy and time. Here it may represent twelve Zodiac ages repeating in precession cycles, and it is the generative/creative source at the center of the six ascending and six descending ages of human mental sensitivity.

How did they locate the Galactic Center?

These data indicate that the Giza monuments plan signifies the location and culmination of the Galactic Center. This should not be dismissed by a hasty assertion that ancient humans "could not have known where the Galactic Center was". How did the monument designers know or measure the location of the Galactic Center? The Galactic Center is currently not a visible sky object.

They could have known its location by four possible ways.

First, the Galactic Center was possibly found essentially by the way we know it. Hershel determined the location of the Galactic Center in the late 18th Century by counting stars and mathematically binning them by magnitude, both naked eye and telescopic stars. The Galactic Center is not a visible object or visible location but the Milky Way Galaxy certainly is visible, though we see it from the inside out. Plausibly, astute and visually acute ancient humans could have deduced a rough Galactic Center location via careful and insightful naked eye observing methods. A non-geocentric conceptual scheme is not even necessary for this method because the Galaxy is a visually apparent object in the sky. But the precision with which the monuments mark the location does suggest a more precise method: possibly naked eye enhancement via telescopes and other observing technologies as we use today.

Second, possibly the Galactic Center was detected by other than visual or astrophysical means. Here, an interesting result from modern studies of anomalous cognition ("AC") suggests a possibility. Of course the study of AC is a nascent science and very controversial, but there is growing acceptance that a weak but statistically robust AC effect does exist. One of the most statistically significant AC effects is the sidereal time correlation stumbled upon by Spottiswoode 1999. Sidereal time is time measured with respect to the "fixed stars" instead of with respect to the Sun. In a statistical analysis of large numbers of AC experiments, Spottiswoode found only one very significant and robust peak in AC, at around 13:30 hrs. sidereal time. Considering the northern hemisphere latitudes of the laboratories where most of those data were taken, 13:30 hrs sidereal time marks essentially a time just after the Galactic Center rises above the horizon. If that AC effect is real, and if human monument designers of 11,000 BC were very sensitive to AC effects, they could possibly have located the Galactic Center (in this case identified only as a sky location with AC influence) precisely without visual observation just by identifying a direction and a time of day.

Third, possibly it was known from indirect information from some other intelligence that used methods 1 or 2.

Fourth, possibly astrophysical activity of the Galactic Center made it temporarily visually obvious. This last possibility is controversial for astrophysicists because dust clouds between us and the Galactic Center are believed to shield us from any visible light that might come from it.

GIZA CONCLUSION

Accurate calculation of Earth celestial pole motions shows that the three Great Pyramids ground plan correspond to the sky plan of the three stars of Orion's Belt in 11,772 BC and 9,420 BC. Also, the ritual star shafts in the Great Pyramid of Khufu simultaneously align with significant stars and with the Galactic Center. The complex of monuments at Giza can be seen as a ground-sky correspondence clock, calibrating the Zodiac Ages to the northern culmination of the Galactic Center in 10,909 BC as the start of the Age of Leo. The evidence for this is strong enough to warrant consideration of implications for understanding other very ancient human-created monuments and cultures. Whether there may be some mechanism by which the apparent motion of the Galactic Center, or even more specific orientations of Sun-Earth-Galactic Center, affects Earth or people, or whether the Giza monument designers marked the Galactic Center for purely symbolic reasons, is not yet determined.

Implications for present and future cyclical cultural effects, signified by the monuments to be related to human cultural capacity, warrant study.

PART IV

Origin Map Discussion

UNIFIED VIEW OF ORIGIN MAP, CALENDAR CIRCLE AND GIZA

Giza summary

The correspondence of the layout of the three Great Pyramids to the three stars of Orion's Belt around 11,000 BC is confirmed. This signification of the southern culmination of Orion's Belt also serves to signify the northern culmination of the Galactic Center. Several other Giza alignments with the Galactic Center and the Galactic Anticenter exist, indicating that the monuments signify the astronomical motions and changing alignments of the Galactic Center, Earth and Sun. The Galactic Center is also marked by simultaneous alignment of the star shafts of the Great Pyramid around 2,360 BC, soon after the nominal date for the construction of the Great Pyramid.

This initial discovery of the galactic alignments and significances in the Giza monuments is astronomically sound and needs further conceptualization and application into standard ideas about human cultural development.

One probable application is the marking of the northern culmination of the Galactic Center as the astronomical calibration point for the start of the Zodiac cycle, which is the precession of the equinoxes, and hence the calibration point for very long-term calendrical systems. Another possible application is an integrated view of the most ancient aspects of the Giza monuments with the Origin Map. This integration seems to involve a system of tracking and of marking specific orientations of the Galactic Center with Earth, Sun and equinoxes.

At the time of vernal equinox sunrise on the date of northern culmination of the Galactic Center 10,909 BC, the Galactic Anticenter was up on the meridian and the Galactic Center was down on the meridian. So the same Great Pyramid shaft that points to the Galactic Center in 2,360 BC aligned near the Galactic Center and Anticenter at geometrical sunrise on 10,909 BC vernal equinox. On the more ancient date and time the shaft nearly aligned through the Earth to the Galactic Center and into the sky to the Galactic Anticenter. This happens every precession cycle, and the variability of the Galactic Center culmination angle is such that on some cycles it is actually at the shaft angle.

These facts raise many questions. How did ancient peoples use this knowledge of such alignments? Why did they monumentalize them in giant architecture? The Origin Map clearly contains more clues to such purposes.

Calendar circle summary

The Nabta Playa calendar circle serves as a star-viewing diagram. It had a time window of applicability from about 6,400 BC to 4,900 BC, when it signified the appearance of Orion's Belt stars as they were seen on the meridian before sunrise on the days around summer solstice. That time window contained the time when the angle of the Orion's Belt constellation was at a minimum, according to its precession motion. The star-viewing diagram of the calendar circle also signified the configuration of Orion's shoulders and head stars as they appeared on the meridian after summer solstice sunset in the centuries around 16,500 BC. This time window contained the time when the angle of Orion's Belt was at a maximum as seen on the meridian during its precession motion.

The star-viewing diagram of the calendar circle shows the user how to easily perceive the motions of precession of the equinoxes by understanding the changing aspects of the constellation of Orion, especially Orion's Belt. And the diagram identifies a time window of attention to the Orion's Belt stars (Alnitak, Alnilam, Mintaka) from 6,400 BC to 4,900 BC, and to the Orion's shoulders and head stars (Betelgeuse, Bellatrix and Meissa) around 16,500 BC.

Origin Map summary and meaning

The Origin Map consists of an extensive system of megalithic structures, and encompasses the calendar circle.

The large megaliths of the Origin Map are in accurately placed alignments that radiate outward from a central point called "Complex Structure A". These alignments point to the same six stars that are signified in the calendar circle star-viewing diagram. And the megaliths align during the same time window as identified in the calendar circle diagram. Also they align to the six Orion stars specifically on vernal equinox heliacal rising. Each alignment is further corroborated by a simultaneous alignment to the brightest star in the north, Vega, which was used as a marker star to corroborate the megalith alignments. *Meaning:* Neolithic humans circa 6,000 BC had a sophisticated understanding of naked-eye astronomy, including the cycles of equinoxes and at least a roughly accurate understanding of the long term precession cycle, and they held such understanding to be important enough to expend considerable resources and social organization to monumentalize the knowledge. Alternatively, the circa 5,000 BC site builders were maintaining much older structures built by peoples who had such understanding.

The distances on the ground to the megaliths in the alignments scale to represent the actual distances in space to the stars. The distances on the ground to megaliths in the corresponding Vega-marked lines scale to represent the actual radial velocities of the stars. *Meaning*: The designers of the megaliths had very advanced knowledge of astronomy, such that we are only able to achieve with modern high technology.

The companion megaliths in each distance alignment match corresponding megaliths in each velocity alignment in such a way as to fit the physics of orbital motion, and thus probably represent actual companion objects to the six primary stars. *Meaning*: The designers of the megaliths had a basic understanding of physics, and knowledge of astronomy that rivaled or surpassed ours today.

The central megalithic structure of the star map portion of the Origin Map contains a sculpture of our Milky Way Galaxy, sculpted onto the bedrock. The Milky Way Galaxy Sculpture accurately represents the shape of the galaxy and the location of our Sun within the galaxy. It is also oriented according to the same system as are the star-aligned megaliths, to the vernal equinox heliacal rising of the Galactic Center around 17,700 BC. *Meaning*: The designers of the sculpture had knowledge of the astrophysical galaxy, and our situation in it, similar to our own such knowledge.

Preliminary evidence also indicates that the other nearby "Complex Structures" contain accurately scaled sculptures of the astrophysically nearby Local Group of galaxies. The meanings of the megalithic structures are summarized in Figure 37. This figure is a copy of Figure 4 with the astronomical identities labeled in gray. The astronomical identities of each of the megaliths are also listed in the table in Appendix B.

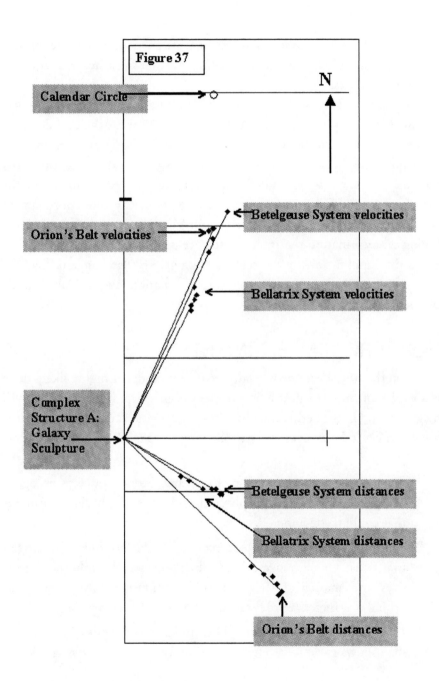

Figure 37

Calendar Circle

N

Betelgeuse System velocities

Orion's Belt velocities

Bellatrix System velocities

Complex Structure A: Galaxy Sculpture

Betelgeuse System distances

Bellatrix System distances

Orion's Belt distances

The different distance scales in the megalith map, used to represent different types of objects, reduce to a scaling law that fits a stepped sequence of prime numbers.

Application of the scaling law indicates that another megalithic sculpture within Complex Structure A is a representation of the cosmic background radiation and hence a reference to the Big Bang of creation. This "Cosmological Sculpture" also has a size, shape and orientation that appears to mark a window defining the whole range of motion of the direction to the Galactic Center, and also to signify one specific time within that window. *Meaning:* The Origin Map may contain valuable cosmological information that modern science is not yet aware of.

Further application of the scaling law analysis also suggests that the Cosmological Sculpture may contain information about Planck Scale physics.

MORE ORIGIN MAP CONTENTS?

Given the finds described so far, it can be considered highly likely there is a wealth of further information in the Origin Map, both in the surface megaliths and in the bedrock sculptures—information that is yet to be analyzed and yet to be excavated. The following is a partial list of possibilities.

• Some of the markings on the Cow Stone, or Cosmological Sculpture, may have astrophysical and/or microphysical meanings that could be deciphered.

• The archaeologists report that many of the surface megaliths are flat oval slabs. This is suggestive of orbital ellipses. Possibly the shapes, sizes and masses of those megaliths contain information about orbital parameters of the stars or planets that they represent.

• In addition to the Sagittarius Dwarf Galaxy, the Milky Way Galaxy map shows galactic structures or possibly more satellite galaxies that are as yet undiscovered because they are hidden from view behind the galactic disk.

- Now that the map pattern can be deciphered, more surface megaliths representing other stars and star-planet systems may be found.

- More galaxy sculptures, on the bedrock, are likely to be found under the other "Complex Structures".

- Probably excavation under the surface star system megaliths will reveal more bedrock sculptures.

- More microphysics information may be found, possibly by continuing the scaling law sequence. More layers of the map system, containing nuclear or atomic scale information, may be present.

- The location of the calendar circle is likely integrated with the stellar systems map. It may represent the precise location of a companion object to our Sun, as is easily calculated using the map deciphering method that I've introduced. If the GPS data for the calendar circle are released, or if its location can be measured, this possibility may be investigated.

Finally, notwithstanding the astrophysical and microphysical importance of the map, the most intriguing aspect of it may be that it appears to describe some sort of function. Such a function seems to involve certain alignments among the supermassive black hole at the Galactic Center, the Big Bang, the Sun (probably at vernal equinox sunrise), Earth's spin axis, and human beings. A unified understanding of this map, together with the alignments indicated in the Giza monuments, and probably with other megalithic structures yet to be deciphered, may pinpoint this function and indicate how to make use of it. This function may involve purely physical processes on one end of a spectrum, or involve human consciousness on the other end of the spectrum. Where on that spectrum of possibilities the actual function exists remains to be determined.

There are hints in some very ancient writings, such as the Sumerian, that might be interpreted as describing such a precise astrophysical-human interaction. Many of Sumerian scholar Zecharia Sitchin's overall conclusions are highly controversial. Scholars seem still to be puzzling over how to decipher the important Sumerian language that predates Egyptian hieroglyphics. But some literally translated Sumerian passages now appear possibly related to the Nabta Origin Map. From a Sumerian tablet that describes "the departure of the two Great Gods", that was planned down to the minute, Sitchin translates: "On the seventeenth day, forty minutes after sunrise, the gate shall be opened before the gods Anu and Antu, bringing to an end their overnight stay." (From Sitchin, *The 12th Planet*.) Such a time window seems strangely correlated with the Galactic Center tracking window in the Cosmological Sculpture that also starts about 40 minutes after rising above the horizon. And on its 17,700 BC vernal equinox alignment date the Galactic Center rose very close with the Sun; thus both bodies entered the "declination window", defined by the Cosmological Sculpture, about 40 minutes after sunrise. Such "galaxy-gates", if they really exist, must entail some mechanism of interaction between human consciousness and physical reality that is as yet unknown.

DATE OF THE ORIGIN MAP

I will discuss briefly three ways to approach the problem of determining when the Origin Map was constructed: geological/stratigraphic; astrophysical; and cultural/historical.

Geological/stratigraphic

Stratigraphically, according to Wendorf et al., most of the star aligned megaliths lie on top of or embedded into a thick layer of playa sediments that was laid down during the "El Nabta/Al Jerar humid interphase". These sediments are radiocarbon dated from around 6,050 BC to 5,300 BC. The radiocarbon dates of cultural materials found generally at Nabta

Playa range from about 10,000 BC to around 3,000 BC with a large number of the dated materials clustered around 6,000 BC. Therefore megaliths on the surface were placed sometime during or after the 6,050 BC to 5,300 BC heavy sedimentation periods.

The existence of the galaxy sculptures cut onto the bedrock, underneath all the sediments, complicates the stratigraphic argument. The archaeologists conclude that after 5,300 BC Nabta peoples created the bedrock sculptures as follows: 1) By some unknown method and for some unknown reason they determined where lumps of bedrock ("quartzitic sandstone lenses") were located under 12 feet of sediments. 2) They then dug a large pit down through the sediments to expose the bedrock, and there sculpted the bedrock lump; 3) They then filled the pit back in with sediments and created the megalithic Complex Structures on top of the backfill.

In light of what those bedrock sculptures really are, another scenario is more plausible. The bedrock sculptures existed from the earliest Neolithic period or before. During the Neolithic period, seasonal residents at Nabta maintained the sculptures periodically and via various methods. When the sediments got too thick on top of the bedrock, surface megaliths were placed. These surface megaliths were intended either as rough facsimiles of the bedrock sculptures or as markers of the bedrock sculpture locations. This clearly occurred for the "Complex Structures", and it is likely to have occurred for the star system aligned megaliths as well. Excavations of the star system megalith sites down to the bedrock is likely to find more bedrock sculptures.

Geologically, an extreme maximum age for the bedrock sculptures is the age of the bedrock. The bedrock underneath Nabta Playa is probably at least several million years old. Some of the oldest bedrock in the region apparently is Cretaceous.

Astrophysical/dynamical

Astrophysically, things move and change. So there are several ways to constrain dynamical ages for the map. A dynamical age tells us when the map accurately corresponded to the moving astrophysical objects, and thus indirectly constrains possible dates of construction of the map. Roughly the same sequence of star alignments occurs every 26,000 years with the precession of the equinoxes. But the precision of simultaneous markings with Vega and vernal equinox heliacal risings together would be significantly in error from the megaliths most cycles around due to ecliptic pole motions and obliquity variations. Therefore the stellar-aligned megaliths currently seen on the surface probably refer to the most recent cycle, because they fit very well. If they are replicas of bedrock sculptures, which may be to the same stars but at slightly variant angles, the detailed alignments of the bedrock sculptures will determine which precession cycle they referred to.

Some of the star alignments are to supergiant stars, like Alnilam, that live only a few million years, therefore giving an age limit of a few million years.

The stars also have relative radial velocities of as much as about 20 km/sec. They are zipping away from our Sun and from each other. In only about 400,000 years some of the relative distances would be about 10 percent out of whack from current values, making the stellar portion of the map inaccurate beyond an age of several hundred thousand years.

Galaxies also move. They give a 10 percent dynamical age limit for the galactic group map of about 500 million years.

The vernal equinox heliacal rising of the Galactic Center marked by the Milky Way Galaxy bedrock sculpture also dynamically constrains its age. The most recent alignment was 17,700 BC. Once each precession cycle a similar alignment occurs. But the motion of the Sun around our Milky Way Galaxy constrains these alignments to roughly a few million years maximum age.

Cultural/historical

Standard theories about human prehistoric cultures may add value to understanding some of the most recent aspects of the map, those megaliths that might have been placed around 5,400 BC. However, standard cultural development theorists would argue that most of the map is impossible. Such theories need to be revised to include these finds, and so they don't add much value to understanding when the map was built.

Age summary

The most recent star aligned megaliths on the surface, such as the Bellatrix-Vega combination circa 5,450 BC, and the one marking the visual heliacal rising of Vega circa 5,400 BC, may have been placed in their locations on top of the playa sediments on the dates of alignment. The older stellar aligned megaliths (the Alnitak, Alnilam, Mintaka – Vega combinations) probably existed before or during the heavier sedimentation period. Thus these megaliths currently on the surface were restored to their current locations from prior settings. This was probably done in an effort to keep the whole pattern of the megalithic complex intact, by ancient inhabitants who sought to maintain the system of structures. The surface megaliths may be replicas of, or fashioned after, bedrock sculptures that predate them; either by several centuries for the most recent alignments, or by as much as 26,000 years.

Some aspects of the bedrock sculptures probably date to at least 17,700 BC, especially the Milky Way Galaxy Sculpture. The extreme age limit for the bedrock sculptures seems to be about half a million years. Based on the available information, my best guess would be the oldest bedrock sculptures date from much later, either about 17,700 BC or about 43,000 BC. But that guess includes a tendency to be conservative about extreme antiquity, a bias that may not be very logical. In some ways it could be more logical if the sculptures are several precession cycles older. Then it is easier to imagine how almost all evidence of the presence of such a highly advanced civilization was lost.

CORROBORATING EVIDENCE

A prehistoric record as extensive and as extraordinary as I've discovered in the Nabta Playa Origin Map is unlikely to exist without any other corroborating evidence. In fact small bits of corroborating evidence do exist. These include but are surely not limited to the following.

Before I was aware of the Nabta site, I submitted a paper titled "Giza Monuments Galactic Zodiac Clock" (Part III of this book) to a major journal. That paper identified that the Galactic Center is signified prominently in the Giza plateau monuments, including the Great Pyramid. That discovery led me to search for important archaeoastronomical evidence in the vicinity of Nabta Playa. So corroborating evidence of some bits of the Origin Map's contents and existence has been present in plain view in the Giza monuments for those who would see it.

Geological erosion analysis of the Great Sphinx sculpture in Giza, by John Anthony West and Robert Schoch, has shown that it is probably several thousand years or more older than historical Old Kingdom Egypt. The Sphinx thus may be contemporary with the Origin Map or with parts of the Origin Map.

Circa 200 BC, Eratosthenes, the librarian at the Great Library of Alexandria, determined the size of the Earth to within 1% accuracy. The children's story told by many scholars goes that Eratosthenes calculated this value from measurements he arranged at Syene, Egypt (modern Aswan) and at Alexandria. These are the latitude and longitude of Nabta Playa. Such a source for geodetic measurements, through the Library of Alexandria repository of ancient knowledge, suggests a possible connection between Eratosthenes and remnants of information from or about the Nabta Playa Origin Map. Eratosthenes also reported very accurate numbers for the obliquity of Earth's axis, and the distance to the Moon. A study of whether Eratosthenes really could have measured these to the accuracy he did, only by the methods attributed to him in the usual stories, or whether he must have used other unknown sources as some scholars conclude, should be instructive. (See e.g. Rawlins, 1982.)

The Piri Re'is 16th century map, investigated by Charles Hapgood and U.S. Navy cartographers, appears to show much of the Earth's coastlines, including the continent of Antarctica accurately. While some wild speculations that the map indicates massive recent "Earth changes" are untenable, the map itself is interesting. That map is a portolano projection centered on a location given by the longitude of Alexandria and the latitude of Syene Egypt on the Tropic of Cancer: this is again the location of Nabta Playa (within about a degree, probably within the standard error ellipse of the Piri Re's map analysis). This suggests that bits of extremely ancient knowledge related to the Nabta Playa Origin Map, and to the people who built the Origin Map, might have survived through history passed by navigational map makers to Piri Re'is. Re'is claimed he made his map as a compilation of other maps; some thousands of years old.

Writer P.D. Ouspensky wrote about his study of the Sphinx: "The Sphinx appears unmistakably to be a relic of another, a very ancient culture, which was possessed of knowledge far greater than ours. *There is a tradition or theory that the Sphinx is a great, complex hieroglyph,* or a book in stone, which contains the whole totality of ancient knowledge, *and reveals itself to the person who can read this strange cipher which is embodied in the forms, correlations and measurements of the different parts of the Sphinx.* This is the famous riddle of the Sphinx, which from the most ancient times so many wise souls have attempted to solve." Ouspensky's teacher was G.I. Gurdjieff, a notorious rascal who dispensed doses of fantasy along with his teachings, and some of whose subsequent students' students now run silly cults. (And many other Gurdjieff followers are very fine productive people.) Gurdjieff emphasized the importance of a "map of pre-sand Egypt", that he claimed to have stolen from a remote central Asian monastery. Gurdjieff viewed his "map of pre-sand Egypt" as central to the development of his esoteric knowledge that did attract some significant followers like Ouspensky. This again indicates that possibly tiny bits of information related to the Origin Map did trickle down through history to Gurdjieff and Ouspensky.

Bauval, Hancock, and Gilbert's realization that the Giza monuments relate to ancient long Pre-Dynastic orientations of the stars of Orion's Belt shows that bits of remnants of information related to the Origin Map survived to today.

Many important people of modern times have been interested in the monuments of Ancient Egypt and considered them to contain important knowledge that has not yet been discovered. These people include founder of modern physics Isaac Newton who spent much time studying the measures of the Great Pyramid, and the founding fathers of the United States, especially Thomas Jefferson who maintained a mockup of the Great Pyramids and Sphinx, complete with sands from Egypt, in the foyer of his Monticello home.

DEVELOPMENT OF THE ORIGIN MAP

The basic enjoyment of deciphering this strange, unusual and wonderful thing has been engrossing. But the importance of the Origin Map site, and the probable future importance of finds there, has not escaped me. In light of the degree of importance and the unique nature of this find, I recommend the following actions regarding the site.

- Egyptian authorities, working with the United Nations antiquities authorities, should move to quickly preserve the site from looters, treasure seekers and uninformed researchers.

- Measure everything and make the measures quickly available so scholars can develop an understanding of the site. Maintain two replicas of the site, one pre-excavation and another post excavation, at another location. This will preserve the look and appearance of the site and its measures, and make them available to researchers.

- Document and preserve everything at the Origin Map site, in light of the extraordinary nature of the find, and also rapidly excavate to the bedrock where much more information is likely to be found.

- Create two international teams to develop and preserve understanding of: 1) the scientific nature of the site; and 2) the cultural-religious-social meaning of the site. These teams should include significant representation from outside the traditional academic communities, as well as representation from the usual orthodox scholars. Make all interim proceedings of these teams rapidly publicly available.

- Initiate a separate effort focusing on the actual functionality signified by the Origin Map. Consider how and whether to attempt to engage a possible coming galaxygate window around the year 2225 (the southern culmination of the Galactic Center), and how to build any structures involved in such an engagement.

- Make publicly available, through internet webcams, constant surveillance of the site. This unusual measure will openly verify the integrity of the site and all future finds there. The potentially extraordinarily important nature of future finds at the site, and the importance of verification as to whether something was actually found there, calls for an unusual public monitoring of this type.

Careful and complete investigation is absolutely necessary, but speed may also be important. Much information about the astrophysical universe, and probably microphysical nature, some of it as yet unknown to modern science, is sculpted in the Origin Map. Careful excavation will reveal and preserve the details. If the Origin Map also proves to contain information on how to build a functional structure to engage a coming galaxygate window, it may take us the 223 years available to do so properly. The steps would involve: 1) secure and investigate the Origin Map; 2) relieve international political tensions enough to be able to work cooperatively on the project; 3) understand the Origin Map information enough to be able to design, build and test a large precise structure in time for the window.

AFTERWORD

by John Anthony West

If you, like me, are among the astronomically challenged, you will not be in a position to pass judgment upon the validity of the evidence Thomas Brophy puts forward to support his thesis. But that thesis is clear enough: the crude egg-shaped stone circle and adjacent stone structures at Nabta Playa in the remote, uninhabited Nubian Sahara enshrine a wealth of complex astronomical information, some of it so sophisticated that only over the past few decades has it become available to contemporary astronomers. If Brophy is correct on all counts, then other, currently undecipherable elements within the Nabta complex may convey information so advanced that even *we* have not yet discovered it. The Nabta Playa formations amount to a kind of cross between an ancient graduate course in advanced cosmology and a planetarium ... but all laid out in rough tombstone-size to car-size slabs across the faceless sediments of an ancient ephemeral lake.

At first glance the subject may seem an arcane, scholarly quibble over ancient astronomical knowledge (or lack of it), interesting perhaps, but with little application to our daily lives; like those heated arguments over how the dinosaurs died. ('Debate Over Dinosaur Extinction Grows Unusually Rancorous' goes my all-time favorite science headline in the New York Times back in the 80's. What is there about dinosaur extinction to get rancorous about?). But if Brophy's and related work is viewed in the context of our modern world and our modern understanding of our own ancient past, then the questions raised by the stones of Nabta take on both personal and global significance.

It is the implications of Brophy's work, and the attendant problems involved in getting a hearing for it that I will address (rather than the science itself) in this Afterword.

In one sense Brophy's work will seem radical; revolutionary. Yet in another it can be seen as just the latest (admittedly most dramatic) contribution to a reappraisal of ancient history that has been lurching along by fits and starts for more than a century—over the raucous, concerted opposition of the entire community of archeologists, historians, Egyptologists, anthropologists and all other academic disciplines devoted to studying the past.

So what is in store for *The Origin Map*? Given the reception accorded far less radical ideas, the reaction to Brophy's claims may be anticipated with some certainty.

There are few things in this world more predictable than the reaction of conventional minds to unconventional ideas. That reaction is always and invariably some combination of contempt, outrage, abuse and derision. As a common corollary, the level of outrage expressed is proportionate to both the quality of the supporting evidence and the magnitude of the challenge posed by the new idea—the better the evidence, the more radical the idea, the louder and shriller the response.

However, this standard reaction may be seriously muted or further enhanced by a potent new wild card, added to the deck only in the latter half of the 20th Century: the PR factor.

If the unconventional idea attracts wide public interest, that is to say, if it is easily understood and 'sexy' enough; and especially if it has resulted in best-selling books, extensive TV coverage or movie blockbusters, the attack gets ratcheted up (on the academic playing field, only the orthodox are allowed to make a living out of their work). In such cases, public acceptance of the offending idea may be successfully diverted or obscured but it will not be quashed. As long as the public interest is there, Hollywood and television can be relied upon to keep stirring the pot no matter what the 'experts' say. And sooner or later the cynics, skeptics and debunkers at the *New York Times*, *Scientific American* and *Skeptical Inquirer* will be forced to confront the offending data. (Defeat is rarely acknowledged 'First they will ignore you, then they will laugh at you, then they will say that everyone has known it all along ' … the attribution is uncertain.) But under enough public pressure, the once-heretical idea gradually takes root and in time may well become a new orthodoxy in its own right.

On the other hand, if the unconventional idea has been restricted to internet dissemination or publication in obscure books or 'alternative' media, the keepers of the paradigm may successfully ignore it altogether or otherwise summarily dismiss it. And it may never get its chance before the jury of public opinion. This approach is made even easier if the evidence for the new idea is so technical that only 'experts' are qualified to pass judgment on it. In such cases, sound, potentially important work may be buried for generations, maybe even for good. There are many instances where derided ideas disappear only to resurface fifty years later, finally recognized as valid, and their original proponents, driven into oblivion, sometimes even suicide by their peers turn out to be heroes two generations later. But no one knows, or can know how much sound, important work still lies buried. My own guess: plenty of it.

The ideas and evidence developed by Thomas Brophy in *The Origin Map* do not fall neatly into either of the categories described above. On the one hand, the ideas have enormous public potential and their implications are, as we shall see, truly revolutionary; on the other hand the supporting evidence is undeniably technical—fully comprehensible only to those with a fairly advanced grasp of astronomy. But it's not difficult to break the problem down into its constituent elements and address each in turn; to explain just what is at stake, and what the pitfalls are to public acceptance.

The ancient view of still more ancient knowledge

Brophy's main idea or thesis (the existence of advanced scientific knowledge in ancient times) is simplicity itself, and it is not new. It would be difficult to find an older idea. Following is a much abbreviated, necessarily over-simplified time-line.

In our schools today we are taught that ancient Greece is the true progenitor of what, after many twists and turns, would become our contemporary Western culture. Everything older than that, or other than that takes a secondary position ranging from merely inferior to primitive and superstitious.

Yet, those ancient Greeks themselves routinely acknowledged their intellectual and cultural debt to the far older Egyptian civilization. (See the extensive exploration of this by Cornell scholar Martin Bernal in *Black Athena: The Afro-Asiatic Roots of Greek Civilization*). Pythagoras is supposed to have studied twenty -two years with the priests of Egypt. And it is through Plato, of course, that we have the familiar account of the downfall of the still-more-ancient Atlantis; a tale related originally, according to Plato, by an Egyptian priest. The Hermetic Tradition, rooted in the conviction of an advanced, spiritual or 'sacred' Egyptian science (long pre-dating the accomplishments of Greece), survived and was devoted to preserving and perpetuating what remained of this ancient doctrine. Hermeticism influenced much of Western and Islamic thought following the decline of Egypt and eventual demise of the Roman Empire.

Despite persecution, the Hermetic Tradition was kept alive through the Dark Ages and the Medieval, largely underground, by generations of Neo-Platonists, magicians, alchemists, Kabbalists and astrologers. In the 14th and 15th centuries it resurfaced, accepted as fact by the scholars of the Italian Renaissance, as they sought to free themselves and all of Europe from the prevailing, stifling religious dogmatism. When Johannes Kepler discovered the laws of planetary motion, he exulted, believing he had *re*-discovered knowledge possessed by ancient Egypt. Isaac Newton, generally regarded as the founder of the modern scientific method, believed (rightly) that ancient measures, existing as long as there had been history, were based upon ancient knowledge of the size of the earth. (Newton also devoted more of his study time to alchemy and Biblical numerology than he did to what would subsequently become modern science—interests scrupulously underplayed by today's historians of science.)

But the history of science, like all other forms of history, is written by the winners. Thus, very few know that in the 17th century, an intellectual battle between supporters of the Hermetic Tradition and those of the school that would eventually produce our rationalistic, materialistic philosophy very nearly fell to the Hermeticists. Had they prevailed, we would most certainly still have an advanced science today, but it would look and 'feel' very different.

The modern view of ancient knowledge

Over the course of the 18th, 19th and 20th centuries, a new institution, call it The Church of Progress, emerged as the dominant religious institution of the Western world. The C. of P. differs from other churches in that it doggedly denies that it is a church at all ('Yes, but I didn't inhale.'). In action, however, it is equally intolerant and unprincipled. And though it denies adherence to a dogma; it has one. Darwinian Absolutism is its philosophical foundation. The fact of evolution through geologic time, as instructed by the fossil record, is undeniable. But Darwinian Absolutism holds to the radical assertion that the mechanism of evolution is only natural selection, a random, purposeless process, necessarily devoid of spiritual or secular meaning. And a fervent belief in the accidental nature and meaninglessness of existence goes hand-in-hand with the equally fervent belief in the invincible and inherently linear nature of Progress itself (Darwinian Absolutism applied to social processes.) It is a given that we, in the 21st Century, are more 'advanced' than any civilization or society preceding us.

Few reading this book have ever been taught, or even had suggested, anything different in any secular Western grade school or university. The long, colorful history of the Western Hermetic tradition goes ignored. And since, in the West, it is axiomatic that objective, scientific truth (the basis of progress) is largely a Western, European development, little credence is given to the traditions of the otherwise highly sophisticated and intellectualized civilizations of China, India, Japan and the rest of Asia, while the beliefs of traditional tribal societies everywhere in the world, (expressed in myth and legend) are given still shorter shrift. Virtually all of these, in one way or another, reference advanced ('Golden Age') civilizations of the past, cataclysmic events that destroyed them, a Deluge and Great Flood.

'Jesuits' of the Church of Progress

The Gospel of Progress is spread mainly through Science, Education and the Press, who operate universally but with varying degrees of intensity and success. Fortunately for the human race, the C. of P. lacks a formal Authority, and therefore obedience to the dogma is impossible to enforce. Though the dreaded CSICOP (Committee for the Scientific Investigation of Claims of the Paranormal) does its best to recreate the atmosphere formerly provided by the Inquisition, heretics can no longer be burnt at the stake considerably compromising its effectiveness.

So it is that while most scientists are enthusiastic devotees of the C. of P. not all are. At the cutting edge of science, an embarrassing minority of the most creative tend to be closet or even outspoken mystics. Thomas Brophy's earlier book was called *The Mechanism Demands a Mysticism*, which speaks for itself. And his title is a quote from Prince Louis de Broglie, one of the most revered names in 20th Century physics.

The Press presents an analogous situation. The hard-core science press is unremittingly C. of P. (see Brophy's account of his attempt to interest *Nature* in his theory) but the mainstream press may accommodate the occasional dissenter. It's not unheard of to find a book anathematized by

Nature or *Scientific American* favorably reviewed in the *New York Times* or *Newsweek*.

With Education it's a bit different. There, Church of Progress dogma is presented unrelieved from kindergarten on. 'Progress' is unquestioned. Darwinian Absolutism is taught as though it were established science. Any attempt to present an alternative view is immediately witch-hunted into oblivion.

Education is not education at all (in the sense of the Latin *educere*: 'to lead forth'). It is a seminary program, designed to perpetuate the priesthood (via graduate studies) while producing a docile and unquestioning laity.

The encouraging thing about that type of education is that it does not work very well. While the young do not have the scientific background needed to see through the sham of Darwinism they often have inquiring minds and open hearts. In practice it proves impossible to inculcate in them an enthusiasm for meaninglessness, and for a philosophy that is, by definition, soulless. Hollywood and Television stand by, ready to fill (and cash in on) the inescapable emotional void created by Church of Progress indoctrination. For half a century the young have been brought up on an 'entertainment' diet of UFO's, alien visitations, reincarnation, life-after-death, faith healing, paranormal powers, high magic and related subjects. It goes without saying that most of this is trash; its sole aim is to make money, not to provide an alternative to rationalist church propaganda, but nevertheless it performs that service in its own chaotic and meretricious way.

One element within that panoply of heresies is the undeniable appeal of 'lost civilizations'; the legendary wisdom of the past; the 'sacred science' of the ancients in general and Egypt in particular. And it is here of course that Thomas Brophy's careful work represents a major contribution—even if only a portion of it should prove valid.

To return to that theme of history written by the winners: the Hermetic tradition and associated ideas had become academically inadmissible by the turn of the 19th century; by mid-century Darwinism applied to 'Progress' made an advanced ancient science academically unthinkable.

Nevertheless, it did not extinguish the interest, and by the latter half of the 19th century, there were stirrings of a renaissance. Here are a few of the major contributions:

Ignatius Donnelly's exhaustively researched *Atlantis* (1889) went into multiple editions and was a huge international bestseller, provoking a spate of books by other authors and very nearly launching a new sub-discipline of 'Atlantology'. In tandem with this a number of highly respected scientists such as Sir Norman Lockyer were beginning to re-evaluate certain ancient structures—particularly Stonehenge and the stone circles and barrows of the ancient Celts—correctly interpreting them as astronomical observatories. Gradually, gradually the evidence started mounting, over the standard objections of the Church of Progress (un)faithful. In 1957 R.A. Schwaller de Lubicz published his massive, 3 volume *Le Temple de l'Homme (The Temple of Man)*, effectively recreating and documenting the sacred science of Egypt and demonstrating unequivocally the reality behind the Hermetic Tradition. In reviewing this book, eminent Egyptologist Etienne Drioton exhorted his colleagues to 'erect a common wall of silence' around Schwaller's work. It took nearly fifty years for this book to finally appear in English translation—a good illustration of the effectiveness of orchestrated academic neglect applied to work that is technical in nature.

In 1968, MIT historians of science Giorgio de Santillana and Hertha von Dechend published *Hamlet's Mill*, proving that detailed astronomical knowledge, including knowledge of the precession of the equinoxes, underpinned the mythology of the ancient world, and even the legends of tribal and traditional people with no apparent current knowledge of astronomy at all—strongly suggesting a common, very ancient source. 1974 saw the publication of Peter Tompkins's best-selling *Secrets of the Great Pyramid* bringing together the best of the 'alternative' pyramid theories and lucidly exposing the utter inadequacy of the accepted 'tombs and tombs only' theory, along with the inadequacy of the standard explanation for how the pyramid was built. By this time the new sub-discipline of archaeoastronomy (the study of ancient astronomy) had taken form. In 1993 our NBC special *Mystery of the Sphinx* brought the water-weathering theory and the need to re-date the Sphinx (and with it all of ancient history) to an initial audience of 30,000,000 and subsequently to multiples of that number. Over these decades, a substantial number of other, solidly researched books and videos added to the accumulating evidence.

In this sense, then, Thomas Brophy's work isn't all that revolutionary; it's an addition to an already considerable body of evidence attesting to a high level of exact science in deep antiquity. It is the height of that level that makes it revolutionary.

Orthodox academics are still vigorously contesting ancient knowledge of precession at all—which requires long, careful (but still only naked eye) astronomy to establish. Brophy claims that extremely complex precessional data has been integrated into the crude, unworked stones of the Nabta formations and also into the plan of the Giza Pyramid complex as well. In very broad terms, Brophy corroborates Robert Bauval's Orion correlation, but interprets the data rather differently, and provides a much richer astronomical explanation for the plan of the entire plateau. He also does not hesitate to suggest extremely old dates (16,000 BC and even older), written into these plans—not necessarily suggesting that's when they were constructed, but rather what they appear to refer to.

Brophy is also claiming that the otherwise inexplicable rows of megaliths near the Nabta stone circle represent the distances from earth of the main stars comprising the constellation of Orion and the speeds at which they are travelling. This is an astounding suggestion! We needed our most advanced telescopes and satellites to acquire this information ... If the distance and velocity of the Orion stars are indeed what this formation is recording, even the most favorable current assessment of ancient knowledge calls for an exponential upgrading. Basically everything we thought we knew about the science of the distant past has to be rethought all over again.

The archeologists who discovered Nabta are in full accord that it was built to enshrine astronomical knowledge; but they were and are unprepared to consider astronomy at this level of sophistication available in 5,000 BC, and possibly even much earlier. Yet, as it stands, without trying to pass judgment upon the validity or otherwise of evidence outside my own field of expertise, what is absolutely clear is that Brophy has provided a consistent, carefully developed and testable explanation for features at Nabta (and at Giza) that could not be explained at all before (e.g. the apparently meaningless megaliths within the Nabta stone calendar circle; the outlying megalith slabs, initially thought to be grave markers, but which, when excavated, produced no artifacts whatever; the apparently random patterns on the peculiar unmistakably hand-worked large stones and areas of bedrock, and much else besides.)

The ball is now in the scientific/academic world's court. Will Brophy's evidence get the careful hearing it deserves? This depends, as I've argued, largely upon the PR factor. If he cannot force that hearing, the chances are it won't happen. Yet, it is a matter of some practical (not merely academic) urgency. Our vaunted Church of Progress has just about brought our home planet to its knees—largely because of its adamant and shrill denial of the sacred. Yet it is very clear that the sacred science of the ancients was real, and that it was science. A validation of Brophy's work would take us one step closer to re-infusing our own brilliant but soulless contemporary science with that ancient sense of the sacred. Without that no fundamental change is possible. And without fundamental change, nothing is possible.

John Anthony West
August 2002

John Anthony West is the leading proponent of the Symbolist school of Egyptology, author of *Serpent in the Sky: the High Wisdom of Ancient Egypt* and creator of the Emmy-award-winning documentary *The Mystery of the Sphinx*. He also personally leads intensive study tours to Egypt every year.

REFERENCES

Badawy, A., 1964, The Stellar Destiny of Pharaoh and the So-Called Air-Shafts of Cheop's Pyramid. *Mitteilungen des Instituts für Orientfoschung, Deutsche Akad. der Wissensch zu Berlin (MIO)* **10**, 189-206.

Bauval, R. and Hancock, G., 1996, *The Message of the Sphinx: A Quest for the Hidden Legacy of Mankind.* New York: Three Rivers Press.

Berger, A.L., 1976, Obliquity and Precession for the Last 5,000,000 Years. *Astron. & Astrophys* **51**, 127-135.

Berger, A.L., 1977, Long-Term Variations of Daily Insolation and Quaternary Climatic Changes. *J. Atmosph. Sci.* **35**, 2362-67.

Electronic Sky, by Mark Fisher. <http://www.glyphweb.com/esky/default.htm?http://www.glyphweb.com/esky/constellations/orionsbelt.html>.

Gantenbrink, R., 1999, <http://www.cheops.org/>.

Gantenbrink, R., 1994, *Mitteilungen des deutschen Archäologischen Instituts Abteilung Kairo (MDAIK)* **50**.

Gilbert, A., 2000, <http://www.adriangilbert.co.uk/docus/articles/star2.html>.

Gillespie, C.C., M. Dewachter, Eds., 1987, *The Monuments of Egypt: The Napoleonic Edition, La Description De L'Egypte.* Old Saybrook, CT: Princeton Architectural Press [orig. 1801].

Gurdjieff, G.I., 1959. *Meetings with Remarkable Men.* UK: Routledge and Kegan Paul.

Hancock, G., and S. Faith, 1998, *Heaven's Mirror.* New York: Crown Publishers.

Heggie, D.C.,1981, Highlights and Problems of Megalithic Astronomy, in *Archaeoastronomy* 3, p S17.

Helmi, A., and White, D.M., 2001, *Simple dynamical models of the Sagittarius dwarf galaxy,* Mon. Not. R. Astron. Soc. 2000, 1-10, December.

Hipparcos: data are easily searched via the SIMBAD database, operated at CDS, Strasbourg, France, <http://simbad.u-strasbg.fr/Simbad#ack-lab>.

Lehner, M., 1997, *The Complete Pyramids,* London: Thames and Hudson Ltd.

Malville, J. McKim, Wendorf, F., Mazar, A.A. &Schild, R., 1998, Megaliths and Neolithic astronomy in southern Egypt. *Nature* **392**, 488-491.

Marriott, C., *SkyMap Pro* Version 6, 1999, <http://www.skymap.com> (Uttoxeter: Thompson Partnership, Uttoxeter).

Mercer, S.A.B., 1946, *The Religion of Ancient Egypt.* London.

NOAA, 2001, <http://www.srrb.noaa.gov/highlights/sunrise/ calcdetails.html>.

Ouspensky, P.D., 1950. *In Search of the Miraculous.* UK: Routledge and Kegan Paul.

Quinn, T.R., S. Tremaine, M. Duncan, 1991, A three million year integration of Earth's orbit, *Astron. J.* **101**, N6, p. 2287-2305.

Rawlins, D., 1982, *Eratosthenes' geodest unraveled : was there a high-accuracy Hellenistic astronomy,* Isis 73, 259-265.

Schaefer, B.E.,1986, Atmospheric extinction effects on stellar alignments. In *Archaeoastronomy (Supplement to Journal for the History of Astronomy)* 10:S32-S42.

Schaefer, B.E., 1993, Astronomy and the limits of vision. *Vistas in Astronomy* 36:311-361.

Schoch, R.M., 1992, Redating the Great Sphinx of Giza, *KMT: A Modern Journal of Ancient Egypt* **3**, N 2.

Sitchin, Z. 1976, *The Twelfth Planet.* New York: Avon Books.

Spence, K., 2000, Ancient Egyptian chronology and the astronomical orientation of pyramids, *Nature* **408**, Nov. 16.

Spottiswoode, S.J.P., and May, E.C. 1999, Anomalous Cognition Effect Size: Dependence on Sidereal Time and Solar Wind Parameters, *Journal of Scientific Exploration* **11**, N2.

Stencel, R., F. Gifford, E. Moron, 1976, Asatronomy and Cosmology at Angkor Wat, *Science* **193**, N4250, pp.281-286.

Touma, J., J. Wisdom, 1994, Evolution of the Earth-Moon System. In *Astron. J.* **108**, N5, p.1943-1961.

Trimble, V., 1964, Astronomical Investigations Concerning the so-called Airshafts of Cheops Pyramid. *Mitteilungen des Instituts für Orientfoschung, Deutsche Akad. der Wissensch zu Berlin (MIO)* **10**, 183-7.

University of Chicago Oriental Institute Giza Mapping Project <http://www-oi.uchicago.edu/OI/AR/98-99/98-99_Giza_fig1.html>.

Wendorf, F. and Schild, R., 1998, Nabta Playa and Its Role in Northeastern African Prehistory. *J. Anthropological Archaeology* **17**, N2, 97-123.

Wendorf, F., Schild. R. and Associates, 2001, *Holocene Settlement of the Egyptian Sahara, Volume I: The Archaeology of Nabta Playa.* New York: Kluwer Academic/Plenum Publishers.

West, J.A., 1993, *Serpent in the Sky.* Wheaton, IL: Quest Books.

Yukteswar, S., 1893, *The Holy Science.* Self-Realization Fellowship ISBN: 0876120516.

APPENDIX A

Calculating Star Locations

Right ascension and declination are found by rotating Earth's celestial pole according to the standard method of Berger (1976, 1979), as follows. Given (x,y,z) a unit vector in the celestial coordinate system of J2000, pointing in the direction of the object, a series of five coordinate rotations gives the future or past Right Ascension and declination of the object:

Rotation about the x axis by angle ε2000, J2000 obliquity, yielding (x', y', z').

Rotation about the J2000 ecliptic pole for the general precession.
$x'' = \cos(\delta 2000) \cos(\lambda 2000 + \Psi \text{date})$, $y'' = \cos(\delta 2000) \sin(\lambda 2000 + \Psi \text{date})$, $z'' = \sin(\delta 2000)$

Rotation about the y'' axis for motion of the ecliptic pole in direction of precession.
$x''' = z'' \cos(\text{Pxrot}) + y'' \sin(\text{Pxrot})$, $y''' = y''$, $z''' = z'' \sin(\text{Pxrot}) + y'' \cos(\text{Pxrot})$

Rotation about the x''' axis for motion of the ecliptic pole in direction of obliquity.
$x'''' = x'''$, $y'''' = y''' \cos(\text{Pyrot}) + z''' \sin(\text{Pyrot})$, $z'''' = y''' \sin(\text{Pyrot}) + z''' \cos(\text{Pyrot})$

Rotation about the x'''' axis by -εdate back to celestial coordinates.

The variables in the rotations are: ε2000 is the obliquity of the ecliptic in year 2000, δ2000 is the ecliptic coordinate system declination of the object in year 2000, λ2000 is the ecliptic coordinate system longitude of the object in year 2000, Ψdate is the angle of general precession from year 2000 to year of date, (Pxrot, Pyrot) are rotation angles for the motion of the ecliptic pole, εdate is the obliquity of the ecliptic in the year of date.

The initial conditions and data values used are as follows. (x,y,z) is a unit vector in the direction of the object, in the celestial coordinate system of year 2000 (from Skymap Pro 6, J2000). ε2000 is 23.437 deg (J2000 from Skymap Pro 6). Ψdate is the angle of general precession (of earth's spin axis with respect to the fixed stars) from year 2000 to year of date. The equations to calculate Ψdate are as shown below. (Pxrot, Pyrot) are the motion of the ecliptic pole with respect to the earth celestial coordinate system from year 2000 to year of date.

Ψdate = Ψbar*T - Δ + $\Sigma_{i=1}^{\text{to } 10}$ A(i) sin[f(i) T + delta(i)], where Δ = 2.12 degrees, Ψbar = 50.439273 arcseconds per year, T = date - 1950 AD, and, delta(i), f(i), and A(i) are:

i	delta(i)	f(i)	A(i)
1	251.9	0.008780547	2.0530611
2	280.83	0.00906125	0.7097639
3	128.3	0.006714499	0.5618778
4	348.1	0.000176866	-0.5482361
5	292.72	0.008884383	0.3445083
6	165.16	0.000871913	0.2649639
7	263.79	0.008603681	-0.2588194
8	15.37	0.012452316	0.2423278
9	58.57	0.000275521	0.1684306
10	40.82	0.000103837	-0.1377861

εdate = $\varepsilon 0$ + (1/3600) $\Sigma_{i=1}{}^{\text{to } 20}$ B(i) cos[g(i) T + deltaB(i)], where $\varepsilon 0$ = 23.32 degrees, deltaB(i), g(i), and B(i) are given by

i	B(i)	deltaB(i)	g(i)
1	-2462.22	251.9	0.008780547
2	-857.32	280.83	0.00906125
3	-629.32	128.3	0.006714499
4	-414.28	292.72	0.008884383
5	-311.76	15.37	0.012452316
6	308.94	263.79	0.008603681
7	-162.55	308.42	0.012130068
8	-116.11	240	0.008957414
9	101.12	222.97	0.008499845
10	-67.69	268.78	0.011855922
11	24.91	316.79	0.012176793
12	22.58	319.6	0.013177622
13	-21.16	143.8	0.017561099
14	-15.65	172.73	0.017841801
15	15.39	28.93	0.000280703
16	14.67	123.59	0.002066048
17	-11.73	20.2	0.01549505
18	10.27	40.82	0.000103837
19	6.49	123.47	0.003671767
20	5.85	155.69	0.017384233

$\text{Pxrot} = \Theta \cos[\text{atan}(\Theta a/\Theta b) + \Psi \text{date} + 90]$

$\text{Pyrot} = \Theta \sin[\text{atan}(\Theta a/\Theta b) + \Psi \text{date} + 90]$ where

$\Theta a = -\Theta a2000 + \Sigma_{i=1}^{\text{to } 15} C(i) \cos[h(i) Tc + \text{deltaC}(i)]$,

$\Theta b = -\Theta b2000 + \Sigma_{i=1}^{\text{to } 15} C(i) \sin[h(i) Tc + \text{deltaC}(i)]$,

$Tc = \text{date(AD)} - 1850$, and $\Theta a2000 = -0.00035$, and $\Theta b2000 = 0.00002$.

i	C(i)	deltaC(i)	h(i)
1	0.01208	12.13810	-0.00156
2	0.00508	305.22144	-0.00188
3	0.02004	249.03308	-0.00523
4	0.00761	277.93504	-0.00495
5	0.02767	106.15321	0
6	0.00281	125.64284	-0.0073
7	-0.00173	316.29332	-0.00083
8	-0.00130	201.28766	-0.00019
9	-0.00025	51.75558	-0.00128
10	-0.00080	313.59074	-0.00183
11	0.00181	265.60396	-0.00215
12	0.00100	237.12278	-0.00505
13	-0.00238	260.94338	-0.00541
14	0.00348	289.84534	-0.00513
15	-0.00076	220.13111	-0.00551

The initial coordinates (J2000 RA and Declination) are either from the SkyMapPro catalog, or the Hipparcos website, except for the Galactic Center which has J2000 RA=17h 45.667m, Dec=29.0078deg. Most of the star locations in the tables in Appendix B also include star proper motions from Hipparcos.

Atmospheric refraction is included in the shaft alignment dates, though the correction is small. The atmospheric refraction used is the standard averaged value, as can be found at the National Oceanic and Atmospheric Headquarters websites (<http://www.srrb.noaa.gov/highlights/sunrise/cal-cdetails.html>).

Figures 38 and 39 show the resulting Earth obliquity, precession and ecliptic pole motions. This celestial pole motion calculation method was verified extensively by comparison to SkyMap Pro throughout SkyMap

Pro's range of applicability (SkyMap Pro uses a different approximation method that is accurate in the time frame 4,700 BC to 8,000 AD, during which it is mathematically essentially identical to the Berger method that is also accurate outside of that time frame), and then the calculation method was applied to the more ancient dates that SkyMap Pro cannot reach.

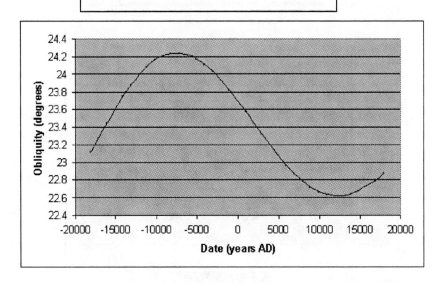

Figure 38: Earth Obliquity and Precession

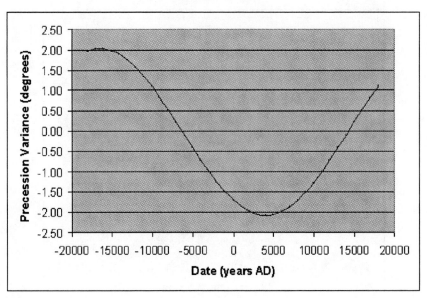

Figure 39: Celestial and Ecliptic Pole Motion

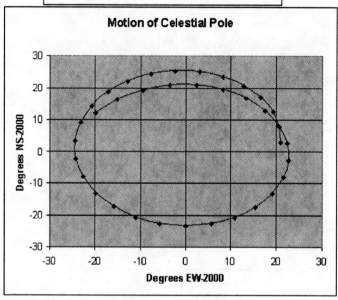

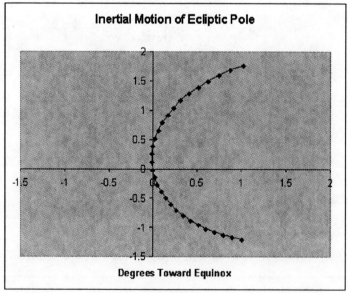

Figure 38 shows obliquity and general precession of Earth as a function of time, over 18,000 years before and after present. The precession variance is the angular difference from the very long term average rate of 360 degrees in 25,694 years. Figure 39 shows the motions of the celestial and ecliptic poles, over 18,000 years before and after present. Knowledge of these three motions, Earth's precession, obliquity and the ecliptic pole, allows the apparent location of a fixed star to be found at any time.

ERROR ANALYSIS

Following the method developed by Schaefer (1986), we can estimate the probability that stellar alignments occur by random chance. This calculation depends on the following values:

σ_A = uncertainty in the extinction angles
σ_D = uncertainty in the date of megalith construction
σ_E = uncertainty in measuring the alignment's azimuth
s = number of stars for which alignments would be allowed
N = total number of available megalith alignments
A = total number of successful alignments on stars.

To make a conservative estimate for the basic megalith alignments, *A* can be taken as 7 (the six Orion stars plus Vega) and *N* can be taken as 10 (six for each of the three southern and northern megalith lines plus four considering the Orion's Belt alignments split into three [note that this is more conservative than just taking *N*=6 because the larger *N* the greater the possibility of chance alignments]). *s* also equals 7, for the six stars designated by the calendar circle plus Vega (which is multiply designated by the diagram itself). σ_D can be considered small (compared to σ_A and σ_E) because the heliacal rising constraint of the system essentially fixes the dates. Then, for ancient Nabta Playa, with a reasonable atmospheric extinction fraction of 0.15 (as noted in the text), and using very conservative estimates according to the method of Schaefer, the ranges of probabilities of random chance become:

star	σ_A	σ_E	single probability	multiple probability
Betelgeuse	0.48	0.5	1.12E-07	6.46E-13
Bellatrix	0.50	0.5	1.31E-07	8.39E-13
Alnitak	0.55	0.5	1.77E-07	1.39E-12
Alnilam	0.55	0.5	1.77E-07	1.39E-12
Mintaka	0.54	0.5	1.67E-07	1.26E-12
Meissa	0.49	0.5	1.16E-07	6.84E-13
Vega	1.01	0.5	2.70E-06	1.30E-10

In the table, "single probability" is the random chance probability that the seven stars align with any seven of ten megalith lines according to the vernal equinox heliacal rising system, using the error estimating values for that star. The "multiple probability" is the random chance probability the system matches repeatedly, six times, two of the seven stars (one Orion star plus Vega).

CORRECTIONS IN THIS EDITION

This printing of the book corrects the following errors that occurred in some copies printed before November 8, 2002. A recursive method of calculating geometric heliacal risings improves the precision of those dates as shown in Figures 6-12 and 17. A minor error in calculating the proper motions of two of the stars is corrected. An improved method of approximating the atmospheric extinction angles determined by Schaefer (1986) is used, and an atmospheric extinction of 0.25 is used rather than 0.15. These corrections are also reflected in Figures 6-12 and described in the section *Heliacal events and atmospheric extinction*. The uncertainty in extinction angles is reflected in the term σA in the random chance probability calculations.

Figure 18 is flipped to show north pole up, correcting a sign error in the galactic pole for that figure.

The corrected dates and coordinates are shown in Table 3 in Appendix B.

Any new analysis such as this is likely to contain some errors of detail. I encourage interested researchers to test these calculations and inform me of further necessary corrections. And I encourage readers to think critically regarding this interpretation of Nabta Playa, remembering that further errors may possibly be found.

APPENDIX B

Data Tables

NABTA PLAYA DATA

Table 1: Stars Coordinates for Calendar Circle.

Star	Declination (deg)	Right Ascension (deg)	Right Ascension (hr min)	Inclination* on meridian (deg)	Date BC
Alnitak	-28.27	1.49	0hr 5.96min	12.88	4,940
Alnilam	-28.06	-0.03	23hr 59.88min	X	4,940
Mintaka	-27.66	-1.53	23hr 53.88min	X	4,940
Betelgeuse	-22.05	-174.38	12hr 22.49min	20.82	16,500
Meisa	-17.53	-177.98	12hr 8.09min	X	16,500
Bellatrix	-19.42	178.21	11hr 52.82min	X	16,500
Sun	24.16	90	6hr 0min	X	4,940
Sun	24.20	90	6hr 0min	X	5,750
Sun	23.37	90	6hr 0min	X	16,500
Sun	22.5083	90	6hr 0min	X	31,330
Alnitak	-34.11	-11.57	23hr 13.72min	13.66	5,970
Alnitak	-36.71	-17.86	22hr 48.56min	14.58	6,440
Alnitak	-37.30	-19.37	22hr 42.53min	14.85	6,550
Alnitak	-1.943	85.190	—	X	J2000
Alnilam	-1.202	84.053	—	X	J2000
Mintaka	-0.299	83.002	—	X	J2000
Betelgeuse	7.315	86.068	—	X	J2000
Meisa	9.934	83.784	—	X	J2000
Bellatrix	6.348	81.387	—	X	J2000

*This is the inclination of the constellation line, not the altitude.

Table 2: Nabta Playa Megaliths.

Line Name	Megalith	Longitude	Latitude	Angle	Star
Complex Structure A	A	30.7256	22.508		The Sun
A3	A-1	30.7299	22.5159	26.70	*Alnilam*
A3	A-2	30.7297	22.5158	25.90	*Alnitak*
A3	A-3	30.7299	22.5155	27.91	*Mintaka*
A3 average				26.83	*Orion's Belt*
A2	A-0	30.7306	22.5165	28.52	*Betelgeuse-A*
A2	A-4	30.7297	22.515	28.42	*Betelgeuse-B*
A2	A-X	30.729	22.5137	28.86	*Betelgeuse*
A2 average				28.60	*Betelgeuse System*
A1	A-5	30.7291	22.5134	30.91	*Bellatrix-A*
A1	A-6	30.729	22.5132	31.13	*Bellatrix-B*
A1	A-7	30.7288	22.513	30.59	*Meissa*
A1	A-8	30.7288	22.5128	31.63	*Bellatrix*
A1	A-9	30.7283	22.512	31.95	*Vega*
A1 average				31.24	*Bellatrix System*
B2	B-1	30.7303	22.5059	115.81	Betelgeuse
B2	B-2	30.7302	22.5059	116.30	Betelgeuse-A
B2	B-3	30.73	22.5061	115.05	Meissa
B2	B-4	30.7298	22.5061	116.09	Betelgeuse-B
B2 average				115.81	Betelgeuse System
B1	B-5	30.7294	22.5061	118.42	Bellatrix-B
B1	B-6	30.7287	22.5064	119.19	Bellatrix-A
B1	B-7	30.7283	22.5066	119.30	Bellatrix
B1 average				118.97	Bellatrix System
C1	C-1	30.733	22.5021	130.80	Alnitak-A
C1	C-2	30.7331	22.5022	129.93	Alnilam
C1	C-3	30.7327	22.5028	128.41	Mintaka
C1	C-4	30.7329	22.5025	129.20	Mintaka-A
C1	C-5	30.7323	22.5029	129.49	Alnilam-A
C1	C-6	30.7317	22.5032	130.42	Alnitak
C1 average				129.71	Orion's Belt

Table 3: Stars Coordinates for Origin Map.

Star	Declination (deg)	Right Ascension (deg)	Date BC
Alnitak	-36.77	-18.00	6,450
Vega	57.81	-148.79	6,450
Alnilam	-35.54	-17.21	6,270
Vega	56.94	-147.39	6,270
Mintaka	-34.07	-16.24	6,080
Vega	56.04	-145.92	6,080
Betelgeuse	-23.31	-10.28	5,820
Vega	54.83	-143.90	5,820
Bellatrix	-24.22	-10.76	5,310
Vega	52.55	-139.96	5,310
Meissa	-19.52	-8.47	5,200
Vega	52.07	-139.11	5,200
Vega	54.92	-144.06	5,840
Galactic Center	-4.95	-2.05	17,700
Galactic N. Pole	4.74	-91.45	17,700
Galactic Center	20.04	90.01	10,909
Galactic S. Pole	-27.18	169.43	10,909

Table 4: Stars Distances and Velocities.

Star	Distance light-years	Distance Error	Velocity (km/sec)	Velocity Error
Alnitak	827	137	18.6	?
Alnilam	1358	370	25.9	1
Mintaka	927	175	16	2
Betelgeuse	433	77	21	1
Bellatrix	246	17	18.2	1
Vega	26	0	-13.9	1
Galactic Center	26,000	1,000		
Andromeda Galaxy	2,400,000	?	-300	?

GIZA DATA

Table 5: Events of 9,420 BC vernal equinox: Orion's Belt stars

Object	Approx Time (24 hrs)	Declination (deg)	Event
Mintaka	23:21	-47.95	Rise behind Sphinx Rear
Alnilam	23:32	-48.74	Rise behind Sphinx Rear
Alnitak	23:43	-49.32	Rise behind Sphinx Rear
Leo	-05:00	~0	Leo rising
Leo	-06:00	~0	Leo lays on horizon facing South
Sun	06:00	~0	Rises in Vernal Equinox
Mintaka	06:00		Orient to ground plan view over Menkaure Temple
Alnilam	06:00		Orient to ground plan view over Menkaure Temple
Alnitak	06:00		Orient to ground plan view over Menkaure Temple
Mintaka	11:30		Set just east of three satellite pyramids
Alnilam	11:31		Set just east of three satellite pyramids
Alnitak	11:34		Set just east of three satellite pyramids

Table 6: Events of 11,772 BC vernal equinox: Orion's Belt stars

Object	Approx Time (24 hrs)	Declination (deg)	Event
Mintaka	00:52	-47.74	Rise behind Sphinx Rear
Alnilam	01:02	-48.88	Rise behind Sphinx Rear
Alnitak	01:12	-49.84	Rise behind Sphinx Rear
3 Belt stars	~00:30		Just above horizon, matching Khufu satellite pyramids orientation to Khufu Causeway.
Leo	~03:00	~0	Leo rising
Leo	~04:00	~0	Leo lays on horizon facing South
Sun	06:00	~0	Rises in Vernal Equinox
Mintaka	04:29		Orient to ground plan view South
Alnilam	04:29		Orient to ground plan view South
Alnitak	04:29		Orient to ground plan view South Culmination
Mintaka	08:30		Set just east of three satellite pyramids
Alnilam	08:29		Set just east of three satellite pyramids
Alnitak	08:29		Set just east of three satellite pyramids

Table 7: Giza Plateau Map Features Measures and Angles

No.*	Feature	Distances Normalized to Khufu-Menkaure distance =1: X	Y	Azimuth from Khufu (deg)	Decli-nation (deg)	Azimuth from Menkaure (deg)
1	Menkaure Temple Center	-0.207	0.783	165.2	-56.9	90.5
2	Menkaure Temple East	-0.227	0.783	163.8	-56.3	90.5
3	Menkaure Causeway East	-0.186	0.783	166.7	-57.4	90.6
4	Menkaure Causeway West	0.468	0.776	211.1	-47.9	90.3
5	Menkaure P 1 East	0.627	0.879	215.5	-44.8	0.0
6	Menkaure P 2 Center	0.681	0.879	217.8	-43.2	205.4
7	Menkaure P 3 West	0.726	0.877	219.6	-41.9	223.0
8	Menkaure	0.632	0.775	219.2	-42.2	0.0
9	Khafre Causeway East detail	-0.384	0.501	142.6	-43.5	74.9
10	Khafre Temple Center	-0.407	0.520	141.9	-43.0	76.2
11	Sphinx Temple South outter	-0.407	0.485	140.0	-41.6	74.4
12	Sphinx Temple South inner	-0.407	0.474	139.3	-41.0	73.8
13	Sphinx Temple Center	-0.407	0.462	138.6	-40.5	73.2
14	Sphinx Temple North inner	-0.407	0.452	138.0	-40.1	72.7
15	Sphinx Temple North outter	-0.407	0.440	137.2	-39.5	72.1
16	Sphinx Temple East	-0.425	0.463	137.4	-39.6	73.5
17	Sphinx Paws	-0.377	0.454	140.3	-41.8	72.3
18	Sphinx Heart	-0.355	0.454	142.0	-43.0	72.0
19	Sphinx Body	-0.332	0.454	143.9	-44.4	71.6
20	Sphinx Rear	-0.303	0.454	146.3	-46.1	71.0
21	Sphinx Rear Enclosure Wall	-0.288	0.454	147.6	-47.0	70.8
22	Sphinx Temple axis-Khafre Causeway	-0.203	0.443	155.4	-51.9	68.3
23	Khafre Causeway West	0.119	0.383	197.3	-55.8	52.6
24	Khafre	0.362	0.372	224.2	-38.4	33.8
25	Khufu Causeway East	-0.890	-0.300	71.3	16.1	54.7
26	Khufu Valley Temple Center	-0.891	-0.330	69.7	17.5	54.0
27	Khufu Causeway West 1	-0.629	-0.117	79.5	9.1	54.7
28	Khufu Causeway West 2	-0.173	-0.008	87.4	2.3	45.8
29	Khufu P 1 South	-0.201	0.151	126.9	-31.3	53.1
30	Khufu P 2 Center	-0.201	0.095	115.2	-21.6	50.7
31	Khufu P 3 North	-0.201	0.035	99.9	-8.6	48.3
32	Khufu	0.000	0.000	0.0	0.0	39.2

* Features 9-22 measures are taken from a detail view of Sphinx area.

Table 8: Alignment Dates, Obliquity, and Precession, Events

Date	General Precession (deg)	Event
13,101 BC	-30.0	Begin Virgo Age, Orion's Belt Set enters map, Orion's Belt Rise enters map, GC rise enters temple
11,772 BC	-11.8	Meridian and Culmination Map Alignment
10,909 BC	0.0	Galactic Center Northern Culmination Marked: Begin Leo Age, Start Zodiac Calibration
9,420 BC	20.3	Sunrise Vernal Equinox Map Alignment
8,707 BC	30.0	End Leo Age, Begin Cancer Age, Orion's Belt Set exits map, Orion's Belt Rise exits map, GC rise exits temple

Table 9: Culmination Dates (GC = Galactic Center)

Object	Date	Type	Dec. (deg)	Precession (deg)
GC	BC 10,909	North	20.04	0.0
GC	AD 2,225	South	-29.05	180.1
GC	AD 14,877	North	15.86	360.0
Alnitak	BC 10,714	South	-50.91	2.67
Alnilam	BC 10,625	South	-50.10	3.88
Mintaka	BC 10,545	South	-49.13	4.97

Table 10: Declinations (in degrees) and Right Ascensions (in degrees)

Date BC:	13,101		11,772		10,909		9420		8,707	
Object	Dec	RA	Dec	RA	Dec	RA	Dec	RA	Dec	RA
Alnitak	-45.85	-133.95	-49.85	-110.29	-50.87	-93.77	-49.32	-65.34	-47.22	-52.63
Alnilam	-44.74	-135.05	-48.88	-111.74	-50.03	-95.44	-48.74	-67.21	-46.75	-54.49
Mintaka	-43.50	-135.91	-47.75	-112.95	-49.01	-96.89	-47.95	-68.92	-46.07	-56.24
GC	16.71	58.62	19.51	77.52	20.04	90.01	18.46	111.38	16.65	121.34
GAC	-16.71	-121.38	-19.51	-102.48	-20.04	-89.99	-18.46	-68.62	-16.65	-58.66
Denebola	-3.28	-36.96	3.31	-20.34	8.04	-9.73	16.49	8.71	20.45	17.83
Omicron Leonis	-26.57	-59.22	-21.70	-40.04	-17.64	-28.32	-9.64	-9.20	-5.59	-0.35
KVT*	-46.29	-128.84	-49.62	-104.68	-50.06	-88.16	-47.58	-60.49	-45.17	-48.32

*KVT is a point in the sky represented by the Khufu Valley Temple, in the sky map oriented and sized to match the Orion's Belt stars.

Table 11: Events of 9,420 BC vernal equinox: Galactic Center

Object	Time (approx)	Event
Galactic Anticenter	Always. (sky-sky relationship.)	Marked by skymap sightline from Khufu (Al Nilam) through Menkaure Temple.
Galactic Anticenter	06:38	Transit meridian at same time as Orion's Belt stars (Mintaka).
Galactic Center	12:44	Rise over Khufu Valley Temple.
Galactic Center	12:44	Seen from Menkaure: rise behind Sphinx rear

Table 12: Events of 10,909 BC vernal equinox: Galactic Center

Object	Time	Event
Galactic Center	11:15	Rise over Khufu Valley Temple.
Galactic Center	11:15	Seen from Menkaure: rise behind Sphinx rear.
Galactic Center	18:00	Northern Culmination meridian transit.
Sun	6:00	Rises in first year of Leo, behind lion rear in sky, due east in front of Sphinx lion on ground.

Appendix C: Photos

NABTA PLAYA PHOTOS

These photos were taken at Nabta Playa, by German journalist Juergen Kroenig.

Photograph © Juergen Kroenig.

The remains of the Nabta Playa calendar circle are seen looking southwest. The two standing stones in the foreground are the east end of the solstice sight line window, and the two standing stones further back and slightly to the left are the west end of the solstice sight line window. The standing stone to the far left is one of the two stones that made the south end of the meridian sight line window; the stones that made the north end of the meridian sight line window are toppled, far right. Inside the circle, two of the six standing stones that made up the star-viewing diagram appear roughly in place (see Part I of this book). The standing stone center-front-left may be the stone representing Betelgeuse; the taller-thinner stone center-back-right may be one of the stones representing Orion's Belt.

Applegate and Zedeno, in the *Holocene Settlement* book by Wendorf et al., pages 463-467, document how Schild and Zedeno used topographical and other analysis to determine the probable initial configuration of the stone slabs when they were intact, as is shown in Figures 1, 2 and 3 of this book, and as was also reported in the *Nature* 1998 paper. They also note that when the archaeologists first discovered the site in 1992, several more of the slabs were still in their intact upright or recumbent positions. Several of the other stones currently seen inside the circle are displaced stones from the outer ring.

Since the archaeological photos published in 1998 (e.g. *Egypt Uncovered* by V. Davies and R. Friedman; and *Nature*), some of the stones have fallen or been knocked over, and some have been displaced by significant distances. A small circular mound of sand around the left-front-center standing stone, in this photo, is probably a result of someone recently propping up the stone; in the 1998 photos these two window stones were closer together. Ongoing degradation of this unsecured treasure is further evidenced by the recent visitors' footprints seen in and around this important calendar circle that is, even by the most conservative estimates, at least 6,000 years old.

Photograph © Juergen Kroenig.

Megalith A-0 was found still standing, sticking about 3 feet out of the sand. Megalith A-0 is the northerly most star-aligned megalith. It is in the line of megaliths that pointed to the star Vega, circa 5,950 BC, simultaneously with the southerly megalith line that marked the vernal equinox heliacal rising of Betelgeuse. See Part II of this book.

Photograph © Juergen Kroenig.

Most of the larger star-aligned megaliths were found toppled and broken but scattered closely around their original locations. The footprints in the sand in the foreground give an idea of scale.

Photograph © Juergen Kroenig.

Many of the megaliths found on top of the "Complex Structures" (see pages 57-60 of this book) were shaped or sculpted by the megalith builders. The footprints in the sand give an idea of the scale.

Photograph © Juergen Kroenig.

Climatologists believe that Nabta Playa, and its surroundings for a hundred miles or more in all directions, has been hyper arid (essentially zero rainfall) for thousands of years. That desiccation and remoteness from inhabited areas, plus the fact that at first glance most of the megaliths look like natural rock outcrops, kept the extremely ancient relics intact, until recently found by archaeologists.

GIZA PHOTOS

These photos of the Giza plateau were taken in June 2001.

The three Great Pyramids are seen from the southwest, looking northeast. Left to right are Khufu, Khafre, Menkaure, and the smaller three Menkaure satellite pyramids. (See the diagrams and discussions in Part III of this book.)

The three Great Pyramids are seen from near the top of the eastern Menkaure satellite pyramid. Left to right are Menkaure, Khafre and Khufu; small and on the horizon to the right of Khufu are the three Khufu satellite pyramids.

The Great Sphinx is seen with the Khufu Pyramid and the small Khufu satellite pyramids in the background, above the Sphinx's shoulder.